Cultivating Change

REGENERATING LAND AND LOVE IN THE AGE OF CLIMATE CRISIS

CARO FEELY

WWW.CAROFEELY.COM

Reviews of Caro Feely's books

Praise for Cultivating Change

'Powerful and inspiring' Jacqui Brown, Book blogger

'A must read. Passionate, challenging, and informative.' Helen Melser, Author

'I have heard it said that time changes things, but sometimes you have to change them yourself. In Cultivating Change Caro Feely confronts both sides of this saying, adapting to the effects of time on family and relationships, while simultaneous seeking to shape the world – especially the wine world – to meet the challenges of the future. Insightful and inspiring.' Mike Veseth, The Wine Economist

'A culinary trip, with layers of biodynamic farming, secrets of nature, environmental activism, family dynamics, and resilience, all tied into a gorgeous package.' Kelly Ryerson, Glyphosate Facts

'I loved this book. It is a fun and enjoyable read, and a brief escape to a special part of the world.' Jeff Harding, Sommelier

'A thoroughly enjoyable read and a wake-up call about how our actions affect our planet.' Tora Shand

'A journey in working with the environment, caring for it and how to incorporate the concepts into our daily lives. Inspirational.' Garry Mason

Praise for Vineyard Confessions (previously Glass Half Full)

'Honest and touching. Caro Feely gives us the real thing including why we need to heal our soil and change the way we farm.' Martin Walker, bestselling author

'Caro Feely is a force of nature! Caro draws the reader into her world with its all of its challenges, triumphs, and heartaches.' Mike Veseth, The Wine Economist

'Vineyard Confessions is a love story poured beautifully onto the pages by Caro Feely. If you love wine or someone who loves wine, you will drink in every page of this book.' Robyn O'Brien, bestselling author

'A brave and compelling tale.' Alice Feiring, author and wine writer

Praise for Saving Our Skins

'Earnest and winning... sincere and passionate' Eric Asimov, New York Times

'So impassioned that it could inspire you to drop all security, move to the backwaters of France, and bet your life, all for the love of making wine.' Alice Feiring, author and wine writer

'Required reading for anyone who loves wine! Even a teetotaller will drink up every page of Saving Our Skins, for the fascinating behind-the-scenes of organic farming.' Kristin Espinasse, French Word a Day

'Caro has produced a beautifully written sequel which in turn seduced and terrified me. I thoroughly enjoyed the urgency of her writing – I needed a rather large glass of wine when I'd finished.' Samantha Brick, author and journalist

'Caro Feely understands that winemaking is an art, a science and a business. Saving Our Skins entertains and informs. Required reading for wine lovers everywhere.' Mike Veseth, The Wine Economist

Praise for Grape Expectations
'Captivating reading for anyone with dreams of living in rural France.' Destination France

'Really liked Caro's book! Definitely the best – and most realistic – tome coming from the 'A Year in Provence' genre.' Joe Duffy, Irish radio personality

'Bright, passionate, inspiring, informative and absolutely delicious' Breadcrumb Reads blog

'Filled with vivid descriptions of delicious wines, great food... a story of passion, dedication, and love' Bookalicious Travel Addict blog

Also By

In the Vineyard Series:
Grape Expectations
Saving Our Skins
Vineyard Confessions (previously titled 'Glass Half Full')
Cultivating Change

Prequel to the Vineyard Series:
Saving Sophia

Non-fiction
Wine, the Essential Guide

Cultivating Change

This 2023 edition published by:

Caro Feely

www.carofeely.com

About Caro Feely

Caro traded in her life as an IT Strategy Consultant to pursue her dreams. She writes books and runs an organic estate with wine school, yoga school, and accommodation, in South-West France, with her partner, Sean. She's an accredited wine educator, a registered yoga teacher, a confident and engaging speaker, and an experienced virtual event facilitator. Follow Caro by joining her newsletter at www.carofeely.com and connecting at www.instagram.com/carofeely or via the social media network links below.

instagram.com/carofeely

facebook.com/caro.feely.wines/

linkedin.com/carow

amazon.com/author/carofeely

goodreads.com/author/show/6176029.Caro_Feely

Note from the Author

This book is memoir. It reflects my recollections of experiences over time. Some names and characteristics have been changed, some events have been compressed or changed, and dialogue has been recreated. Thank you for joining me on this journey.

We are all connected on this shared planet, every one of us a potential butterfly wing for another being's hurricane.

I dedicate this work to you, to us, to the changes we must embrace, and the joy and healing they will bring.

Contents

CHAPTER 1

Seduction in Saussignac

A photo caught my eye as I flicked the pages of a magazine. Deboussey was hunched over, older than when we met. He looked weighed down. Decades before, he dipped a pipette into a wine-stained barrel, and transferred a dream into my outstretched glass. Now, his dark, curly hair, was salt and pepper, his eyes had lost their sparkle.

In another lifetime, in a cellar of rough limestone, stained red from centuries of nurturing wine, we tasted his vinous art. He charted season, and harvest, his shepherding of it from pruning to bottle. The tasting set Sean and me on a quest to become winegrowers, on a transformation from city professionals, to organic farmers. I read the article with a sick feeling. Deboussey had lost his vineyards in a forced liquidation sale. His wine journey was over.

Dread flooded through me. With each passing season we experienced more extreme weather events, wider swings of temperature. Farmers lost their crops to hard late frosts, hail, drought. The climate crisis was taking its toll across the globe. While the article did not put his situation down to climate change, I knew his region had been hit by frost, and hail, in the preceding years.

To ease the stress headache I felt tightening across my forehead, I reminded myself that winegrowing could be generous, as it could be cruel, like each season. We loved our farm. Its renaissance was thrilling; wild orchids flowered in abundance and large populations of beneficial insects buzzed through our fields. I needed to focus on the positive, and get on with making dinner. I put the magazine aside.

The lentils felt edgy in my hand, like my mood. I threw them into a pot, and turned to look out of the window. Nature had painted fifty shades of green across the landscape. Below the house, the rows of vines running down the hill were covered with lacy new growth. Across the courtyard, the barns took on a honey glow of early evening, contrasting with the oak trees' bright spring foliage, and a cerulean blue sky. I took a long, slow breath, and felt my anxiety drain away.

A shadow came into view, followed by a muscular body, topped with a trim, white beard and bald head. Adrian, or Ad for short, waved, and a few steps later, walked through the front door. His hand lifted to his mouth, his eyes widened, and his eyebrows shot up.

'Caro, you won't believe what happened today.'

'Tell me,' I said, eager for something to take my mind off Deboussey's story.

'I was standing at this window, and a beautiful, young girl flashed me!'

Ad was working on a welding project with Sean. I wondered if the heat had got to him. He saw I was incredulous.

'Caro, I promise you. It happened right here.'

He moved in next to me, to act out the scene.

'I came in to check my emails on your Wi-Fi. I looked up, and there she was, flicking her skirt up, and wearing no undies.'

His eyes popped at the memory.

I laughed at the craziness of it. Was it hallucination from overwork? Or were sirens practicing nude seduction in our courtyard? Ad was super fit. He could cycle to Bergerac without breaking a sweat. He, and his wife Lijda, had dropped in the year of our first grape harvest. We needed help and they offered it. Over those intense days, they became like family. From then on, their campervan parked up for a week in autumn, and sometimes in spring. Ad and Lijda became surrogate grandparents for our daughters, filling in for their real grandparents who lived far away.

Now, a decade later, Ad was retired and sometimes visited on his own. He was one of several exceptional people who had helped us transform the farm from a failing venture – it was in liquidation when we bought it – into a viable business. For Sean and me, the transformation had come at the cost of neglecting other aspects of our lives. It was our passion, but, at times, it risked overwhelming us, generating burnout, or breakup.

Ad left, and I turned the pot down to simmer. Sean stomped in and squeezed into the entrance alcove to take off his boots. Spring growth meant long days of physical work in the vineyard, on foot, or in the tractor. He preferred to be on foot, but both left him tired and hungry by the end of the day. He looked up, his bearded face framed by a checked shirt and frowned.

'Not lentils again,' he said.

'It's not lentils, it's dhal, spicy lentils,' I replied defensively.

Lentils were a staple for me. I loved them simply as lentils and rice boiled up together, served topped with pesto and yoghurt, and a side of salad. Tonight, I was exploring wider horizons with the dhal. Sean and I took turns to cook. He put more effort into dinner on his night than I did on mine. He had a point, but I wasn't going to concede it. Sean climbed the stairs to shower and I returned to the bubbling pot.

A half-hour later, we sat down to dinner and Ad entertained us with his story of the dancing nude. Sean and our daughters, Sophia and Ellie, were as mystified as I was. We debated who it could have been. I had welcomed a wine tour group that day. But why would one of them have flashed our friend? It was a mystery.

'So I finished the trailer,' said Ad, changing the subject. 'I think I'll ride into Ste Foy tomorrow, then head home on Friday.'

'Thanks Ad, the trailer will be a big help,' said Sean.

The converted flatbed trailer would bring buckets of handpicked whole-bunch grapes from the vineyard to the winery. We had changed from machine harvesting to handpicking a couple of years before. It brought a different spirit. Handpicking was joyful and intense, rather than panicked and noisy. The selection and care of handpicking translated into better quality wines. Harvest remained a moment of pressure, the year's work coming to fruition, but it felt vibrant, rather than crushing. Ad's harvest trailer meant Sean could collect more buckets with one trip, and he had re-modelled equipment we already had, but no longer used.

'We'll miss you Ad,' I said.

After Ad left, Sean and I plunged into the season, siren forgotten. From April to October, we were both stretched with vine growth and visitors. Sean was trimming the vineyard by hand for the first time. It took time, but was worth it. Machine-trimmed vines are neat green stripes. They have lost their glorious freedom. Swirling canes and curling tendrils waving in the air are part of a vine's natural way of being. I love them like that, bright green dancers on the soft breeze.

In biodynamic farming we consider the vine's year cycle like a mini lifecycle. In winter, when the vines drop their leaves and go to sleep, it's like death. Spring is like a rebirth. The growth phase that creates canes and leaves is like a child growing up. When the vine flowers in preparation for fruiting, it transform from a growing adolescent into a reproducing adult. Her focus moves from leaf growth to grape growth. If we hedge trim the vine we trap her in adolescence. She is unable to make the transition to adulthood. She will replace the leaves and canes she lost instead of focusing on her fruit. The hormones that manage the change from growth of canes and leaves, to maturing the fruit, are controlled in the vine tips. Constant trimming of the tips removes this ability. Each cut also offers a potential entrance for infection.

Biodynamic farming is like organic plus. Organic farming means no chemical fertiliser, weedkiller, systemic fungicides, or systemic insecticides. 'Systemic' means the pesticides go inside the plant and cannot be washed off. Organic allows 'contact' sprays only – sprays that are not systemic and can be washed off – and that are naturally occurring. Biodynamics is a step beyond that. It is stricter in terms of the rules we must follow, and also philosophical. In biodynamics we think of the farm as a whole farm system, a self-sufficient entity, almost like a living thing, an eco-system. We use teas and biodynamic preparations to reinforce the health and vitality of the farm. We also follow the biodynamic calendar, a lunar calendar, like the farmer's almanac. This calendar helps us to do things at the best time for the vines and the wines.

A few weeks after Ad left, the vines flowered under sunny skies, promising a bountiful harvest. The flowers transformed into baby grapes that were like peas, hard and green. Over the next couple of months, they would grow and soften, filling with the sun's energy transformed into sugar by the vine's superpower, photosynthesis.

I walked along the Cabernet Sauvignon, stopping intermittently to check the grapes' development. Further on, on the plateau near Saussignac, our local village, I did the same with the Merlot and the Sauvignon Blanc. The grapes looked healthy, on track for their seasonal development.

Sophia, Ellie, and I, were on our way to Gageac-et-Rouillac, a nearby village, to attend a rock concert, the '*Festival des Ploucs*'. The festival for 'country bumpkins', the loose translation of *ploucs*, was a highlight in the calendar. The concert rotated through the four villages that made up the Saussignac wine area, or appellation, a different one each year. It attracted thousands of visitors.

The winding path followed a hedgerow from Feely farm, past our neighbour, to Saussignac village. We descended into woodland below Saussignac Castle, the air thick with forest fragrance and excitement. I dropped back, slowing down to enjoy the dappled light and allow the exuberant youngsters to get ahead. The space grew until their voices were a distant murmur. I closed my eyes and inhaled the aromas. Peace settled around me, and the trees came alive with birdsong. My body filled with well-being.

After a few long breaths, I took off again, slowly closing the gap. My tribe's laughter rose in cadence with the increasing volume of the distant music. We popped out of the forest onto a track surrounded by vineyards. An avenue of ancient mulberry trees led to the imposing 16th century Château de Gageac, its 10th century *donjon* tower visible above a fortified wall.

Fields transformed into parking areas gradually filled with cars. A row of tents lined a hedgerow, ready for overnight revelers. Inside the festival boundary in the shadow of the castle, young and old thronged in groups, chatting, or waiting in loose queues, for tickets, or food and drink. Beyond them, near the stage, a small crowd danced and swayed. In a couple of hours, it would be a thick mass.

My teenage daughters took off to join their friends. We had attended this concert since they were in pre-school. Being part of a community where it felt safe to let my daughters disappear into a concert crowd was a gift. Gratitude washed over me. I made a tour to say hello, bouncing from friend, to neighbour, to acquaintance.

I stopped to chat with a colleague I had not seen in a while.

'How's the vineyard doing?' I asked.

'Ah Caro, I'm not a winegrower anymore,' replied Sandrine wistfully. 'I've changed my profession. I'm training to be a nurse.'

'That's a big change.'

'It is. It's been hell, but I'm starting to see the other side.'

'I can imagine. Changing your job while you're raising a young family. That's hard. Why did you do it?' I asked.

'It's a long story, but the short version is that we got divorced.'

'I'm so sorry,' I responded.

'I know. We lost our harvest to frost a couple of years ago. The financial stress was too much. I'm not saying it was the only thing, there were cracks in the relationship already, but it didn't help. Anyway, I decided to change direction. Being a winegrower wasn't something I wanted to do on my own. It's a demanding job, and getting harder.'

'Tell me about it,' I said. 'You're brave. That's a big transformation for you. How are the girls taking it?'

'Oh, you know, they've adapted. They have a week with me, a week with their *papa*. It's part of their routine now.'

This was a climate story. It wasn't one that we saw in the news, like floods, or fires, but it was fallout from the climate crisis. With climate change came extremes, including harsh late frosts, that could wipe out a harvest.

There is a double effect of climate change that contributes to increased frost risk. Temperatures are hotter earlier, so the vines set their buds earlier when there's still a high chance of frost. Then, there are more temperature swings, ups and downs, not only more heat.

Three years later, in April, after the vines had budded, we experienced a temperature swing from minus 2 degrees Celsius in the morning to plus 22 degrees Celsius in the afternoon, a massive 24-degree difference, where usually, the swing from high to low was closer to 10 degrees Celsius. We lost about 20 percent of our crop to that frost. Some growers lost 100 percent.

When the temperature goes below zero Celsius, new vine growth is frozen and dies. We pray that another shoot will grow, but it probably won't be a fruiting shoot. Even if it is, the fruit is unlikely to reach useful ripeness.

The climate crisis also creates reinforcing negatives. Frost means that farmers burn fuel, run fans, and even fly helicopters overhead, to save their crops, thus reinforcing the original problem by generating more carbon dioxide. The same happens when we run air conditioning on a fossil-fuel-based electric grid. We reinforce the problem and create a negative, reinforcing cycle.

At the time that I bumped into Sandrine, I was filled with hope that governments were going to make the necessary changes required to solve the climate crisis. COP (Conference of Parties) 21 had recently taken place in Paris. It ended with a historic consensus by representatives of the 196 attending parties (195 states and the EU). In the final adopted version of the Paris Agreement, the parties said they would pursue efforts to limit the temperature increase to 1.5 degrees Celsius. Keeping warming below 2 degrees Celsius was thought to be critical to avoiding tipping points that could destabilise the earth's life support systems in a way that would be devastating for us humans and for many other living things.

'How are you finding your new career?' I asked.

'It's hard work, but I enjoy it. I miss being in the vines, but at least I have a stable job. Nurses are always in demand. I won't lose everything to fickle weather.' She half smiled and shrugged her shoulders.

We chatted for a while, then said affectionate farewells. A small stab of adrenalin flashed through me, similar to what I felt after the Deboussey liquidation story. Sandrine was tough. She was transforming a bad situation into something positive, but her family unit was broken, and she had scars from the difficult years of transition.

As the sun dropped towards the horizon, I found a seat next to Véronique and Sébastien Bouché, the couple who had designed and built our tasting room, and the Wine Lodge accommodation, a few years before. They were a key part of the story in my second book 'Saving Our Skins'. Since then, Véronique and Sébastien had created their organic farm with glamping called *'Ecolieu Cablanc'*, a kilometre from Saussignac village. We shared a passion for ecology and organic farming, and had children of similar ages.

Red wine and conversation flowed. Véronique had initiated *'Incroyables Comestibles'*, 'Incredible Edibles', for the village and school. These were small food gardens in raised beds, for people to contribute to, and to use. She said it had been a bit of a one-way street. People had not participated as much as she expected. We talked about how to raise the ecological vibe in Saussignac. Our village was a *territoire bio engagé*, an engaged organic territory. This meant at least 35 percent of the arable land was farmed organically, pesticides were not used in shared outdoor spaces, and 35 percent of the food served in the school canteen was organic. Despite this progress, we knew we could do better.

'Oh, but tell me Caro, what's going on with Château Feely?' asked Véronique, her red hair flaming in the setting sun and her eyes bright. 'Who are these man-eaters you've been entertaining?'

'What?'

'You don't know what happened?' Véronique raised her eyebrows, and glanced at Sébastien. I shook my head. 'Sébastien was over at your place, creating the Cottage fence for you.'

'Ah, yes, I remember the project. What about it?'

'You don't know!' She drew in her breath dramatically. 'Sébastien was hit on. I mean hit on, in a big way.'

I looked from Véronique to Sébastien and back to Véronique. Now my eyebrows were raised.

'Yes! Sébastien was working, and a young American girl came over and started chatting him up. She came on to him, hot to do it right there.'

'No way!'

'Yes! And she was serious. I know Sébastien is good looking, look at his great body, his blond ponytail.' She paused. Sébastien angled his head in a model stance and gave a broad smile to gently mock her compliment. We giggled.

'But to hit on him like that, in broad daylight, on a family farm!' continued Véronique, more animated with every word. 'The brazen girl! She said to Sébastien, 'Let's get it on! Come on! Let's hop in your van and do it right here.' Sébastien showed her his ring and said, 'I'm married'. She said, 'your wife won't know'. But of course,

I would know. I look into Sébastien's eyes, and I know everything.'

Véronique gave me a wink. Sébastien laughed, part flattered, part shocked, at the memory of being '*dragué*' – seduced – by a twenty-year old. Véronique's blue eyes flashed.

'That wasn't all. When her word-based seduction didn't work, she turned around and flashed him! Lifted her skirt and showed him her nude bum. She wasn't even wearing undies! Flicking her skirt up in your courtyard, like it was The *Moulin Rouge*. Can you imagine?'

I was laughing out loud. The penny had dropped.

'This explains our friend Ad's experience! He was in our kitchen and a girl flashed him from the courtyard. We couldn't understand it. This explains the mystery. He was witnessing the seduction of Sébastien.'

'No Caro. No. The 'attempted' seduction of Sébastien,' said Véronique, laughing with me. 'It's truly unbelievable. That hussy! 'I'm here, you're here, a quickie in the van,' she said. Sébastien phoned me and I told him, 'Come home right now! If you don't, I will come over and commit murder.'

She paused, then leant in conspiratorially.

'I'm a very jealous partner.'

We all laughed.

'I can't believe how forward these young Americans are. Imagine doing that!' She leant in close again. 'Caro, I don't think Sébastien can work at Château Feely anymore.'

I laughed heartily. After a few more stories, and the last drop of our shared bottle of red, I went to look for Sean. He had arrived to work the last shift in the winegrower tent, a rustic bar where wine from Saussignac was sold by the glass, and by the bottle.

We were like ships passing in the night, seldom doing things together. I found him, wished him and the rest of the winegrower team *bon courage*, then rounded up our daughters. It was the right moment to leave, the peak of the evening before the descent into drunkenness. Sean would see some unpleasant scenes before he got home in the early hours of the morning. Wine, like any alcohol, required moderation, responsibility, and respect.

The warm night was like a cosy embrace. At the stream in the nook of the valley, before the sweep up to Saussignac, an owl whooshed down through the trees. We stopped. The sound of owl wings cutting air joined the distant sounds of the last band warming up.

I had a sense of community that night. We had lived long in three countries: South Africa, Ireland, and France. We were part of all three, their essence bound in ours. Sean and I grew up in South Africa. We both had Irish heritage and a work opportunity took us to Ireland in our mid-twenties. A year after moving to Dublin, on our first wine holiday, we met Deboussey, and fell in love with France, and the dream of being wine farmers. Seven years later, we made the leap to our farm in Saussignac, the story told in my first book, 'Grape Expectations'.

We would never have the same experience as someone who had grown up in Southwest France. Cultural references like French TV shows, personalities, or music, from eras before our arrival, passed over our heads, as did complicated wordplay, but we were settling.

While I was starting to feel a sense of belonging in France, I missed the connection of close family and old friends, people that knew me deeply from long ago. We were forging links, but it took time. Connections to friends and family are important, they are the fabric of our lives. They give our lives meaning. Part of our current attachment to material things is an effort to make up for what we are missing in connection in our lives. I had much to learn on the subject. Remedies were on the way, a lifelong friend was about to visit, and a trip to South Africa was planned. We would introduce our daughters to our original homeland and connect to our roots for the first time in more than fifteen years.

CHAPTER 2

Belonging

A slim figure stepped off the train at Gardonne station, a few minutes' drive from our farm, and about an hour from Bordeaux city. She tossed her bag to the man next to her and ran towards me. I ran too. We clung to each other. It had been too long.

'Now that's what I call good friends,' said the bag carrier.

'Caro; meet Dave,' said Steph as we pulled apart, her tear-stained face alight with a wide, familiar smile.

Dave put the bags down and gave me a hug. I felt like I knew him even though we had just met. He loved Steph and she loved me. He instantly took me into his circle. They were golden brown and sparking with energy. Standing on the platform, I was filled with sensations of my first home in South Africa. They carried it with them, in their accents, mannerisms, vibrancy, and the bright colours they wore.

Steph and I forged our friendship in a small-town preschool. Now we lived on different continents and hardly saw each other, but when we did, it felt like what Aboriginal Australians call the 'everywhen'. Time was meaningless and we flowed, our atoms in sync.

We had two days to introduce them to Saussignac, running, laughing, filling the air with shared memories, and catching up on the time passed. Friends are part of what makes life worth living. Xavier Rudd, Australian singer, songwriter, musician, and activist, wrote and performs a song called 'Walk Away' in which he talks about leaving what is not working for you, and knowing yourself. One lyric is 'my dear friend, so nice to have you home', perhaps about welcoming himself home. For me, it resonated with memories of homecoming, of connections that went beyond needing to see one another regularly, of feeling at home, regardless of the place. That's how it was with Steph.

It wasn't long before we saw them again. A few months later, we flew to South Africa. In Durban, the air slapped us like a hot facecloth, thick and soupy with humidity. At Zinkwazi, forty minutes north of King Shaka international airport, Steph and Dave were waiting on the road, torches in hand, their shadows like sprites in the car lights.

We parked and hugged tight. Up a grassy hill, their beach shack welcomed us. I grew up in KwaZulu Natal, the province where we landed. There was a familiarity in it, the sounds, the feeling of the air. For our daughters, it was a culture and temperature shock. They were born in Ireland and lived most of their lives in France. They had never seen wild monkeys jumping freely in the trees, or geckos on walls, or sensed humidity like this on their skin.

We shared a light snack then hit our beds. The travel had taken its toll. As I fell asleep, something stroked my cheek and it wasn't Sean, snoring at my side. I considered getting up to turn on the light, but my body wouldn't move. Sleep took me and with it my thought 'Was that a moth or a gecko?'

The next day I woke with eyes cushion puffy. Long distance air travel wasn't good for my health or the planet's. Sean was preparing coffee. His stable force and morning routine offered an anchor to our daughters in this new environment. I stumbled to our Robinson Crusoe bathroom. Wood slats surrounded a large unglazed window that looked out over a tropical canopy to the Indian Ocean shimmering in the distance. The light was so bright it felt like midday, but the rented mobile phone said 7.30 a.m. Steph had said they ran at seven. I was too late. I pulled on my shoes, eager to see the place anyway.

The cottage was built from reclaimed wood and second-hand material. Dave had picked up the windows, walls, kitchen cupboards, and doors, from dump trailers outside building sites in Durban's wealthy suburbs. Love and effort had transformed what was tossed away, into a home.

Excited voices pitched across the terrace next to our Crusoe shower room. I followed the sound. Two small boys were playing in a tool shed alongside the terrace. I introduced myself, and gave them a double hug. It was the first time I had met my friend's sons; they were asleep when we arrived the night before. I knew them. I had seen photos, but it was more than that. They were extensions of my friend. I knew them deeply like I knew the vibration of their DNA, not only from photos.

'Where are your mum and dad?' I asked.

'They went for a run,' said Picasso, the older one, responsible at six years of age.

'Are you guys okay?'

'Sure,' said Picasso. Jethro looked up and nodded.

'Okay, see you later. I'm going for a run too.'

I waved to Sean and our daughters taking their coffee on the balcony then let myself out of the garden gate, homemade from recycled metal and wood. A creeper had grown over it, and turned it into a living sculpture. I ran down to the road below.

'Ha ha ha da da, ha ha ha,' cried the hadada birds, an ibis native to Sub-Saharan Africa, large and raucous, named for its loud three to four note calls. They sounded like they were laughing, and it took me back to my childhood serenaded by them as I climbed trees in the garden.

Exhilarated, I followed the road down to the sea. Zinkwazi was unknown and yet familiar. Rich aromas of tropical plants, fertile earth, and sea, mingled. The smells and sounds were still a part of me despite three decades of living in another hemisphere. I felt like a time traveller. As I wound my way back, more hadadas laughed. I laughed with them, feeling freedom, happiness, and a shot of endorphins, as I climbed back up the hill.

Steph and I prepared breakfast while Dave organised the family for a morning on the beach. Sean, Sophia, and Ellie joined us to organise the table. Bananas were set beside ProNutro, a maize and soya cereal unique to South Africa, and oats. The ProNutro ingredients list said the maize and soya were genetically modified. I reached for a banana and topped it with oats. At home, walnuts graced the table. Here, it was locally grown pecan and macadamia nuts. We finished eating, and Sean took Sophia and Ellie to get ready for the beach. Dave's mum looked in the door, and Steph introduced me. Jane and Khulu, Dave's parents, lived in an old brick house a few metres away.

'Look how white she is,' said Jane. 'Not like us, nut brown Africans.'

I was moon white, like a night moth emerging from the European winter, blinking in the harsh African sun.

'Is Khulu here this afternoon? Can he take us to Lala Forest?' asked Dave.

'I'll check with him,' said Jane.

She headed off.

'Jane was a doctor until she was forced to retire at eighty,' said Dave. 'She received a letter saying she had to stop, that GPs were expected to retire at sixty-five. She would have kept working. Now she's teaching the local scouts; and Khulu's got a part time job as a conservation advisor that gets him out of the house a couple of days a week.'

'I love that they're still actively contributing to the local community in their eighties,' I said.

Everything here was larger than life, even the grannies were made of steel.

That afternoon Khulu knocked on the glass front door to offer his taxi service to Lala Forest. He was lean, a ball of energy at 84. We followed his swift, agile lead up the steep slope to the main road behind the house. Conversation came easily.

'I'm a naturalist. Into conservation, fauna and flora, ecology.'

'Just don't give him a gift or a card that isn't homemade,' said Dave.

'Why buy it when you can make it?' asked Khulu.

We jumped into the bakkie, the name for a small pick-up truck, or utility vehicle in South Africa, me in the front with Khulu, everyone else on the back. The sandy track, hugged by sugar cane, felt like another world. It helped to take my mind off our business in France. The upcoming bottling, booking enquiries and accounts, could wait. Being chained to our business was a choice. I pushed the thoughts away, and admonished myself, I had the power to define my work life balance.

The pick-up bucked along the track. Sophia and Ellie gripped onto the bars on the back, eyes nervous, but excited. Rolling cane fields, wild tropical forest and ocean, filled the windows. Far below us, white necklace waves unfurled onto golden sand. A couple of hours before, we were on that beach swimming, fishing, and walking. Now we were off on a forest walk. Steph and Dave's household was constant action. Swimming, walking, running, eating, talking. Sophia and Ellie were acclimatising. Africa was bright, brash, surging with energy, edgy, the opposite of sedate Europe.

Khulu dropped us between two walls of tropical forest. Dave led us to a path into the green growth.

'Lala Forest,' said Dave. 'A ravine so deep, the local tribes hid one thousand cattle here to save them from Shaka, king of the Zulus.'

Thick, thorn encrusted vines dropped down through trees, palms, and ferns. Strelitzias, that looked like big banana trees, nestled against Natal Mahogany, a large, evergreen tree with dark, glossy leaves. It felt like a Jurassic Park film set. We wound down, down, down.

'Do you remember the route?' asked Steph.

'Not really,' said Dave. 'It could be that way, but I think it's this way.'

Like I said, Africa was edgy. There was a chance it was this way. Or perhaps not. We were deep in the forest. We had no water or phones.

'Relax C,' whispered Sean, picking up my anxiety.

We continued for a while, hoping Dave had selected the right route. The vegetation thinned and the ground flattened.

'Ja, this is it. We're on the right track,' said Dave.

'Isn't this where Khulu said the eagle's nest is?' asked Steph. 'Is that it?'

'Yes! That's the fish eagle,' said Dave. 'Well spotted.'

A few metres away, haphazard branches stacked high in a giant Natal Mahogany, gave it away. We moved closer.

Ellie was a keen photographer. Using the magnification of her camera, she showed us the eaglet, nearly as big as a laying hen. At full size, females have a wingspan of 2.4 metres. The bird watched us back.

'They're part of this place,' said Steph. 'Zinkwazi' means 'place of the Fish Eagle'.'

'Khulu keeps an eye on them as part of his conservation work,' added Dave. 'They're an indicator species. If they go, conservationists know something's wrong.'

'I've heard of that. In Kenya fewer fish eagles tipped the authorities off to systemic pesticide run off from flower farms and coffee farms[1],' I said.

'Crazy to think that buying flowers, or coffee, can harm fish eagles,' said Steph.

Sophia, Ellie, Jethro, and Picasso, had stopped listening. They were lost in the wonder of the living forest and the mighty bird above us. Ellie snapped a last photo and we took off again. The sun was setting as we dropped out of the forest onto the beach. Ghost crabs played on the wet sand between each wave. We walked home, cool sea soothing our bare feet.

The next morning, I was up to see the sun rising over the ocean, serenaded by a cacophony of birdsong.

'Gwree tree tree treee'

'Wrrrrr, wrrrrrr, wrrrrrrr'

'Bring me, bring me, bring me'

'Matoe dipi stutu'

'Ha ha ha da da, ha ha ha'

The air was warm and petal soft. I sat on the rustic veranda waiting for movement in the main shack. '*Shayile*' was hand painted over the entrance of our cabin, finished a day before our visit. It meant 'knock off time'. A metre-high heart-shaped mosaic enclosed the word 'home' on the wall near the front door. The rustic interior was filled with turquoise fabrics and handmade objects. It felt like holidays. The house had no alarm, no burglar bars.

'It's safer to have nothing worth stealing than to try to create Fort Knox,' said Dave.

After hearing sounds in the main house, I let myself in and started unpacking cutlery from the dishwasher. Above the countertop, a panoramic window looked out onto the glittering ocean. It was 7 a.m. and the glare was so bright I had to put my sunglasses on.

'Hey Caro!' greeted Dave. 'Ready to run? Khulu's going to drop us so we can run near the river.'

'Sounds groovy,' I replied.

Steph came out of their bedroom, just woken, and hugged me. 'It's so good to have you here.'

'Thank you, my friend,' I replied, eyes filling with emotion.

Soon, we were thumping along in the bakkie, me in front with Khulu, Dave and Steph in the back. Khulu stopped to point out a family of vervet monkeys in a roadside tree. Tiny newborns clung to their mums.

'They're so cute,' I said, marveling at their faces, bright with curiosity.

'Yes, but you must be careful. They go into houses and take anything left out. They're smart,' warned Khulu.

We turned off the tar road onto a sandy track.

'Khulu, tell me about the plants in Lala Forest,' I said.

'Oh, there are ferns, palms, and fruit and nut trees. A key tree is the Natal Mahogany,' he said pointing to one beside the track. 'It's what local people used for oil, to moisturise their skin, and to care for leather, before modern commercial oils. They got the oil from crushing the seeds.'

Dave banged on the top of the cab.

'Khulu, this looks good,' he shouted and leapt off before we had come to a stop. We climbed out, and joined him and Steph, beside a hedgerow that ran along a sugar cane field.

'Caro, this is our energy food when we run. Try one,' said Steph. She plucked a small, shiny, orange-pink fruit, the size of a date, and passed it to me.

I bit into the skin, and with a little pop, it gave way to sweet, juicy flesh and seeds. It was delicious, easy to eat on the run, and tasted like it was packed with vitamin C.

'I wonder why it isn't commercialised?' I asked.

'I don't know,' replied Dave. 'Maybe it won't grow intensively.'

'It has a tannic bite, perhaps it's too tannic for modern palates,' I said, taking another one. The tannin didn't put me off. I loved it. Double bagging my morning cup of tea, and tannic red wine, were my thing. 'Or maybe it's too delicate for modern food logistics.'

Modern, intensive food production means that our food supply has decreased in biodiversity. Even within a specific fruit, for example the apple, the number of varieties has decreased significantly in the last 100 years. In the early 1900s, around 14000 varieties of apples were grown in the USA. In 2020, that was down to 2500, with around 100 grown commercially, and three varieties representing more than 50 percent of the market. Eating wild, as we were doing, was a way to diversify our diets. Buying or growing heritage fruits and vegetables, that are not standard supermarket fare, is another way to do it. Biodiversity reduces risk. The principle 'don't put all your eggs in one basket' also applies to food. Dan Saladino explores this idea in his book 'Eating to Extinction'. He outlines that of the 6000 plant species humans have eaten over time, just nine provide 75 percent of colories consumed by us today.

'Perhaps we should go a little further,' said Khulu. 'The river is still a long way away.'

I nodded enthusiastically, and I reached for another orange-pink wonder.

'Okay, a little further,' agreed Dave reluctantly.

'The fruit is called '*amathingulu*'. It's a bush that's indigenous to this area,' said Khulu as he picked up pace again.

Another bang on the roof stopped us. Dave had found a good place to start our run. Khulu left us, his pick-up truck kicking up a puff of dust as it took off. We followed a sandy track towards the river. The sweet, grassy aroma of sugar cane rose on the warming air. We crossed the river, then passed through more cane fields. A half-hour later, a band of tropical forest took us onto the dunes near Zinkwazi village.

'We'll get the canoes and go down to the lagoon,' said Dave, as we ran the last stretch along the beach.

Within minutes of getting home, the bakkie was packed, and a group of us were jammed in the back alongside five canoes. We flew down the palm-lined road, hanging to the sides like limpets. Sophia laughed wildly next to me.

'I can't believe we're doing this. My friends would never believe it,' she shouted.

The light was so bright, the green so vivid. Another truck flew passed. There were no protection bars on their pick-up either. One of the passengers was sitting on the back corner.

'That would be SO illegal in Europe,' shouted Sophia.

Sophia wasn't usually keen to be outdoors, or to take physical challenges, thanks to me being an over-protective mother. She had a dangerous start to life, that left me risk-averse, and cautious for her. Major surgery at less than 24 hours old was a recipe for new parent stress. My book 'Saving Sophia' charts the story. After her cliffhanger start, I didn't allow her to take normal childhood risks. In South Africa, despite her cautious mother, she was raring to go, keen to do things she hadn't done before.

Picasso took his canoe off the pick-up, raced down the grass bank, and floated it onto the lagoon. Then he tied his brother's canoe behind his and they took off.

Sophia pointed to them and gave me a look that said, 'How can I possibly match that?' She clambered into her canoe and headed off undeterred. Soon her elegant figure was on the other side. I mounted my floating steed and caught up.

'Let's go upriver,' I said.

'Yeah!' she responded.

When we got back, we stepped into deep mud at the lagoon edge.

'The sign of a good day out,' I said pointing to the muddy streaks on our calves.

'Yuck,' responded Sophia.

As we rinsed the sludge off our legs, Steph, Sean, and Ellie arrived.

'How about a turn on the canoe?' I asked.

'I don't want to go,' said Ellie.

'You'll love it,' I said.

'I don't want to go,' repeated Ellie, wondering if her mother was deaf.

With a little persuasion she was paddling. A couple of hours later it was time to go home for lunch.

'I don't want to stop,' said Ellie.

This trip was an opportunity for us to widen our world, like Ellie was doing with the paddling, but also to connect with old friends. Steph had invited a dear friend to join us. He arrived soon after we got back to the house.

'Yo, Caro! So good to see you!'

Russ hugged me and I felt like I had when Steph arrived at Gardonne train station. Russ and I were close in our teenage years and had travelled part of Europe and New Zealand together in our early twenties. We had been buddies since pre-school.

We walked up for ice creams at the corner store, catching up along the way. Life's roller-coaster ride, family, lovers, children, friends. I went to pay for the ice creams.

'Let me get that,' said Russ.

'No, I will,' I said.

He ignored me and paid. We sat under parasols licking the ice creams and talking non-stop. Russ filled in gaps in his life story. He had recently married the love of his life. They had created a new business together and put all their energy and savings into it. The year before, after six years of building it, the business folded. Big supplier companies had agreed to be part of it at first. But once they realised that their margins were being slashed by the grouping of individual orders, they pulled out and destroyed the business. Large companies had the power.

Russ had found a job and was reconstructing his life. I wanted to rewind so I could pay for the ice creams. He had always been generous. No matter what was happening in his life, he was always the first to offer to help.

'But tell me about you, Caro,' said Russ.

'We celebrated our ten-year anniversary on the farm last year,' I replied.

'I didn't realise you had been in France that long. I feel like we're still teenagers.'

I looked across at our daughters, real teenagers.

'Me too. You're living proof that it's about attitude, not years,' I said, punching his shoulder lightly.

He laughed.

'What's happening with your book talks? I saw something online. I want to come.'

'Thanks, Russ! I'd love that. My book distributor has organised events to coincide with our itinerary. I want the next ten years to include more sharing on organic farming and environment. The events are a toe in the water. I'm so chuffed you want to be there. Thank you.' Chuffed was slang for 'delighted', often used in South Africa.

'I wouldn't miss it.'

Back at the house, mouth-watering smells wafted through the kitchen. Sean took our daughters down to collect shells on the beach. He was considerate, giving me time with my friends without being asked. Gratitude filled my soul. He knew Steph, Russ, and I had a special friendship. I didn't know how important that time was.

'It's almost done,' said Steph.

'I feel like a walk on the beach. Will it be okay if we leave it?' I asked.

'Absolutely.'

'Let's go!' exclaimed Russ, his energy and enthusiasm filling the room.

The wind was picking up, but it didn't put us off.

'Your boys are magic Steph,' said Russ.

'Thanks Russ. I was a career girl when I met Dave. Having kids wasn't on my radar at all. But when we had the car accident, my perspective changed. I lost half my hand and part of my wrist, but I was alive. Suddenly nothing seemed as important as having a family. I was already step-mum to Dave's sons from his first marriage, but I wanted my own. Our boys are my life, but it hasn't stopped me dreaming and developing ideas for a business. We visited India last year and I met incredible women who make cloth from organic cotton. I would love to do something with that. But when we got back, I realised it was impossible to fit in being a mum, running our lives, and starting a business. I want to enjoy my sons. I don't have time to start a business right now.'

Sand stung our legs, and the wind almost blew us off the beach, but we kept walking. Steph was right. What is more important than family and friends in the end? All the rest is ash in your hand. Tolkien's great adventure 'The Hobbit' carries the underlying message that friendship is more valuable than gold and jewels. A Harvard study on happiness[2] that ran for almost 80 years found close relationships were what kept people happy throughout their lives, much more than money or fame. Relationships protected people from pain, from life's discontents, and even helped delay mental and physical decline. They were better predictors of long and happy lives than social class, IQ, or genes.

'You're making a good choice,' I reflected. 'There'll be time for a business later if you want it.'

'I agree,' said Russ. 'Starting a business takes time, that's for sure. I learned a lot creating our business.'

'And there's opportunity behind each challenge as Foo always says,' I added.

'She's wise. How's Foo?' asked Russ. Both knew my sister, Jacquie Somerville, aka Foo, well.

'She's thriving. She's a life coach and entrepreneur in Los Angeles. She loves it there.'

'That sounds so Jacquie.'

'How about your new job Russ? What's it like?' asked Steph.

'It's with an NGO in agriculture. I'm fired up about it. The only thing is it's based mostly in KwaZulu Natal. That means Mel and I are apart for three weeks out of four because I'm here rather than home in Johannesburg. The positive side is that I see more of my family down here.'

'It must be tough,' I said.

'We do online date nights when I'm away. She's an amazing woman. You'll love her. I can't wait for you to meet her.'

'Hey why don't we meet in Italy to celebrate our 50th birthdays? Steph and I have been talking about it for years. How about you and Mel join us?'

'It's still a little way off for us,' replied Russ pointing to himself and Steph, ragging me, the oldest of the three. We laughed.

'Let's do it,' said Steph. 'It would be so good.'

'I'm in.' Russ gave us his wide smile that made it feel like everything was possible.

We turned around and were pushed along the beach by the wind, racing back to the house where our families gathered.

In the kitchen Dave was setting wine glasses out. He took a Springfield 'Life from Stone' Sauvignon Blanc out of the fridge, an iconic South African wine.

'Wow Russ! Only the best hey!'

'Ah Dave, mostly I buy the ones I know, the ones I have visited. I feel like I know them better if I have been to their tasting room. There's some insider info for you Caro.'

'So our tasting room and tourism activities should pay off more than we think?' I replied with a wink. 'But where's the organic wine?'

'Look! I got one that's organic and no sulphites,' said Russ holding up a Stellar Organics 'no sulphite' Merlot.

'Now we're talking,' I replied.

Dave poured glasses of Springfield. It was zesty and bright like I remembered. We had visited the estate decades before when we lived in Cape Town. It nestles in the Robertson Valley, on the Breede River, surrounded by a ring of majestic mountains. Their 'Life from Stone' was reminiscent of a Loire Valley Sancerre Sauvignon Blanc and Feely *Sincérité*, packed with flinty minerality, fresh cut grass, lime, gooseberry, and a touch of passion fruit.

'Can I try the red now too?' I asked grabbing another glass. 'I want to taste it before I start eating and my palate is changed.'

'Sure. *Ag ja*. I forgot. We're with the professionals now,' replied Dave.

I too had briefly forgotten. With a glass in hand, every wine was up for analysis, not merely a drink to be shared. The no-sulphite Merlot was pure berry, like a fresh bowl of blackberries, with a hint of mint.

'I love it,' I said. 'I'll drink that with dinner.'

'I can't believe it,' said Russ, tasting it in turn. 'It was the cheapest wine of the four bottles. I bought it as the joker in the pack. I expected it to be horrible. But that's darn delicious.'

Sean, Sophia, and Ellie returned from their walk and joined us for drinks, listening to the conversation with the tight circle of friends I had known long before they were part of my life. As we swapped stories on the terrace overlooking the sea I felt deep contentment, the kind you get when you are surrounded by those you love.

Sean was stepping in, looking after our daughters so that I could have time with my pals. He was thoughtful. But I felt frustrated in our relationship. We didn't make time for the two of us. We assumed our emotional partnership didn't need nurturing. Since moving to France, our business and parenting partnerships were the only ones we invested in.

The next morning Dave, Steph, Russ, and I, went for a run. Our conversation turned to organic agriculture.

'I'm sceptical about organic's ability to feed billions of people,' said Dave. 'I think the future is in genetic modification.'

'I agree,' replied Russ.

'No way you guys,' I exclaimed, shocked. 'Genetic modification is playing God with the potential for scary human error.'

'What do you mean?' asked Russ.

'There can be unintended consequences. Changing one gene can set off other changes. People also make mistakes. In 2009 three of Monsanto's genetically modified maize varieties failed. Hundreds of thousands of hectares of South African maize were barren[3]. No harvest. The company paid out more than forty million US dollars to compensate the farmers for their lost crops[4].'

'I never heard of that,' said Russ.

'Me neither,' said Dave.

'The compensated farmers were subject to a gag order and couldn't talk to the media or researchers about it. It's almost impossible to find information on it. They squashed it.'

'Sounds sinister,' said Steph.

I nodded and raised my eyebrows. 'Another reason I don't like genetic modification and manufactured patented seeds is that farmers lose their ability to be self-sufficient and to save seeds that have adapted to local conditions.'

I tripped on a stone and fell onto another, cutting my hand slightly. I bounced back up and got back to the details, eager to shed light on the issue, particularly given their work. The NGO (non-governmental organisation) Russ worked for was a joint venture that included an agricultural chemical company that manufactured pesticides and genetically modified (GM) seeds. They were working to bring GM seed into South Africa. Dave was a lecturer in agriculture at a Durban university.

'It's also about resilience. I could go on.'

'Go on then,' replied Dave, laughing.

'Agriculture based on factory seeds that need to be bought every year makes no sense. Gathering seeds that have adapted to the local microclimate, courtesy of nature, is a much better idea.' I stopped talking to catch my breath, then continued. 'Nature is adapting to climate change. We need to use her magic. Plants and animals are moving poleward at more than four metres a day. They are onto it and making changes.'

'Amazing Caro,' said Steph.

We arrived back at the house. The run was over and I felt like I still had convincing to do. I was happy Russ was coming to my book event. We'd get to catch up there again.

'Adventure!' I exclaimed holding my wounded hand up when we got home.

'Euh! Are you okay?' asked Sophia.

'Yes. A pesky rock. I had my glasses off because they had misted up in the heat.'

'There were two rocks on the entire sandy track, the one that you tripped on, and the one that cut your hand,' added Dave.

'You need to be more careful Caro,' admonished Sean.

'I need to lift my feet, and get back to running like a gazelle,' I said, winking at him. In our twenties he used to say that about me.

Dave disinfected my hand. He picked up good practices from his life of adventure. He survived sharks tapping his surfboard, being shot by a robber in their house, their traumatic car accident, and a serious slashing by ropes while out kite surfing. He stared death down and came out the other side graced with energy that exceeded most of us. Now he did his best to avoid robbers and car accidents, but he hadn't been put off surfing and kite surfing.

As we wended our way towards Howick, to stay with Sean's dad, I felt an almost physical pain I missed Steph so much. I had no idea when we would see her and her tribe again. The seed of meeting to celebrate our 50th in a couple of years had been planted, but it was a vague plan.

CHAPTER 3

Speak your truth

My friend Russ, and a group of people from the district where I grew up, attended my book event in the city of Pietermaritzburg. I shared stories of our experience with organic farming and life in France. Before leaving the bookstore, I bought the latest edition of a magazine that was a regular read for me when we lived in South Africa.

Inside, an article about the benefits of genetic modification (GM) in agriculture was laid out alongside a full-page advert for Monsanto GM seed. The ad was presented as an infographic that looked like part of an article. It said GM generated better food supply and better health.

I felt compelled to write to the editor. I explained that GM could lead to total crop failure as was the case with the failed GM maize, which was worse food supply. Also, with GM, the farmer was fully dependent on seed companies.

This made GM worse for food supply and resilience. GM Roundup Ready crops meant farms were spraying more herbicide[5] than ever. Roundup Ready seeds germinate plants that are resistant to the weedkiller Roundup and can be sprayed with it. These plants will likely have residues of glyphosate since they can be sprayed directly with the chemical. Glyphosate weedkiller, which includes the brand of weedkiller 'Roundup', made by Monsanto, now part of Bayer, was recognised as a probable carcinogen for humans by the World Health Organisation the year before[6]. Use of glyphosate had recently reached about 100 times the level it was in its first decade of use. To me, this signified worse health, not better.

That evening, as we sipped on a delicious Reyneke organic Shiraz, found at a local wine shop, Sean and I did a post-event review.

'It was good,' said Sean. 'But maybe you went too strong against herbicides and pesticides given we know some of the attendees use them.'

'I have to say it like it is. I can't change the truth,' I said.

I knew I had to speak out given what I knew. But perhaps Sean was right. Did I turn people off by coming on too strong? The following day, some of the attendees shared messages including 'interesting and good food for thought'. It was confirmation that I had to speak my truth.

Dad Feely lent me a book by Antjie Krog called 'Country of my Skull'. Antjie was a journalist who followed the Truth and Reconciliation Commission, a restorative justice body set up at the end of apartheid. She is a mighty writer, poet, and long-time activist despite her roots as a Free State Afrikaner from a farming family that was deeply conservative. I was gripped by the book and her courageous exploration of terrible and beautiful moments through the journey of the Commission's work as they sought to create forgiveness and understanding in unforgiveable situations.

In high school I read 'A Dry White Season' by Andre Brink and was shocked. Apartheid was horrific and its application was brutal and terrifying. People who spoke out against it died mysteriously. Before reading that book, banned when it was first released, I had no idea we were living under such violence. In apartheid South Africa, the media was strictly censored.

Sean was an activist at university, forcing political change in the late 1980s. He was a member of the ANC (African National Congress), a banned organisation at the time. He participated in demonstrations against apartheid. I didn't have the guts to demonstrate, to risk prison, or worse, my life. Sean did. He stood up and spoke his truth. He cultivated change.

When Nelson Mandela was freed, I had recently started my first full-time job working as a business analyst in Cape Town. The Sunday of his release, I walked out of the hostel building where I was night warden, onto Hope Street, and down towards the city centre. Crowds had gathered to await his arrival. The city echoed with joy and expectation. It was electric. We were at a critical turning point for South Africa. I wanted to go out and be part of that crowd, but I hadn't earned my place in it. I hadn't actively participated in the struggle. I was a silent supporter and that didn't count.

Reading Krog's memoir helped me understand more about the country of my birth, its transition, and the ferociously brave people that stood up and sometimes even gave their lives in the struggle against apartheid.

I met Sean a couple of years after Mandela's release. He was a journalist covering the heady race towards South Africa's first democratic elections. We were skint, in the thick of paying off our student loans. We dreamed of going on a luxury safari but the closest we got was camping in our 2-man tent at a game reserve in Eswatini, a small independent country that borders South Africa and Mozambique. Sean planned to propose to me there, but I got chatting with some Aussies and soon we were part of a campsite party, Sean's romantic two-some scuppered.

Sean's next proposal plan was a weekend in Cape Town, more specifically, on the bench dedicated to his winegrowing grandparents, on Helderberg Mountain. We climbed the hill and, as we reached the bench, a storm cloud opened. I sprinted for the car, leaving Sean alone, dripping, and, unknown to me, engagement ring in hand.

I had found the ring on an antique market months before. The aquamarine blue stone and gold filigree caught my eye, and I knew it was the one. We had been looking for a while and none of the manufactured engagement rings excited me. When I saw this one, I immediately put down a deposit so the seller would hold it for us. I gave Sean the details, then heard nothing. In the intervening months I sabotaged his proposals in Eswatini and Cape Town.

Shortly after the Cape Town weekend my brother Garth dropped in. He asked about our wedding plans.

'It's taking longer than I expected,' I said. 'We talked about getting married but there's no date set and no

official engagement yet. I chose a ring months ago but
since then nothing. I'm starting to wonder.'

'Caro, guys will never get married if they're not
pressured to. You must put the screws on. You're not
getting any younger. If he leaves you, you'll be an old maid.
You need to get that ring.'

I was twenty-six and surprised by his view of the
situation. But my brother knew more about being a guy
than I did. By dinner, I was convinced Sean had no
intention of getting married. Two steps in the door after
a rough day I attacked him with the accusation and burst
into tears. Sean calmly requested more information about
what was going on. Then he hugged me and assured me
he had the ring. He wanted it to be a surprise. I asked
forgiveness for doubting him and pulled him closer.

Sean and I had lived together for a couple of years
by then. The legal protection offered by marriage was
not available to unmarried couples, as it is through
common-law marriage legislation widely available today.
Garth was trying to protect me.

It was also societal. Women were raised to see getting
married as their destiny. I was a career girl and considered
myself a feminist, but I was influenced by that pressure. I
believed I had to get married, that my life was somehow
not complete if I wasn't. I wanted that ring.

A couple of weeks later Sean invited me to pack my bags
for the weekend. 'And don't forget your passport.'

We drove north of Johannesburg, in the direction of
Lanseria airport where small planes destined for private
game reserves took off. I felt butterflies in my stomach.
Perhaps this was the luxury safari I had dreamed of. Sean
kept driving past the offramp and gave me a wide smile. A

couple of hours later, we followed a dirt track to a country hotel with a large wraparound veranda. The 'don't forget your passport' was a red herring but I was more than happy. Sean knew me well. He knew I would prefer that he paid off his student loan.

We dropped our bags in a rustic room, the smell of thatch mingling with fresh linen, then wandered down the dusty track that had brought us in, eager to see our surroundings before nightfall. Loose limbed and free of the city, we felt at ease. The smell of gum trees and dry grass, the sounds of crickets winding up for their night-time chorus, and of birds chattering as they bedded down, filled my soul. I reached for Sean's hand.

It was two years since we met on an Economics study weekend, and I still felt fizzy and lovestruck when I held his hand. That evening, a glass of Cap Classique sparkling wine arrived, and I looked for the ring at the bottom of the glass but came up empty. By the end of the sparkling, my expectations were forgotten. I was deep in the luxury of time together, leather chairs, linen tablecloths, a faint smell of cigar smoke and aromas of cooking herbs, cinnamon, and old wood.

Our conversation flowed with the courses, from a starter of springbok pâté served with homemade bread, to a main course of guineafowl, butternut mash and spinach, paired with a luscious Shiraz. We talked of our dream of wine farming. I felt the bliss that came with food, wine, and dreams, shared with a lover. After the cheese course, we were invited to take port on the veranda. I sank into the worn leather sofa and set my glass on a low table. A zebra skin's dramatic black and white stripes peeked out from under it. Sean put his glass beside mine. The murmur of the restaurant hushed. The air was warm, stars flung across the sky, a bright streak of the Milky Way down the centre. I felt time stand still.

Sean took my hand and dropped to his knee.

'Caro, will you marry me?' he asked.

Tears rushed down my face.

'I thought you wanted to get married. Are you okay?'

'Yes. I'm crying because I'm happy. I love you.'

He slipped the ring onto my finger, the semi-precious aquamarine stone set between inter-twined leaves of gold, sparkled in the candlelight. To the sounds of the bushveld, Sean recounted my sabotage of his proposal campaign. It was third time lucky. We laughed quietly then blew out the candle and found our way to our room.

With time to study the ring more closely, I realised it was a grape and vine leaves. Our future was written in the stars, and even in the ring I chose.

Now, two decades, two daughters, and a vineyard later, on our first family trip to South Africa, we were getting a luxury camping safari. We told Sophia and Ellie we were going camping, leaving out the luxury part. They hated camping. When they saw the luxury canvas with a view over a waterhole, stone outdoor shower, deep bath, and lounge and bedrooms like something out of a Condé Nast magazine, they were relieved and thrilled.

We saw more big game than I had ever seen, elephants, lions, buffalo, even a lonely cheetah, the last one left on the reserve. At one point, we were so close to a lioness, I could have almost touched her. But something jarred with the mad chase in 4x4s. At times it felt like canned photo hunting for the big five, rather than a deep wildlife experience. Game watching where you sit quietly at a water hole and listen, take in the vibe and sounds of nature, offers a more soulful way. In a city, finding a quiet spot near a canal, the sea, or in a park, can provide this nourishing connection with nature.

The ranch had been a reserve for about a decade. It was filled with big game and their food. They had four of the big five – lion, elephant, buffalo and rhinoceros – necessary to attract clients and make it economically viable as a safari destination, but it was missing the massive biodiversity of a long-term reserve like the Kruger National Park, with its nearly two million hectares – about the size of Wales – established in 1898. As we had discovered on our farm in France, land that has been abused takes time to heal. The night sounds in the Kruger Park are rich with biodiversity, small mammals, insects, amphibians, a great web of life. It would take decades to reach a similar richness of life.

We travelled on to the Cape Province, host to an entire plant kingdom called the Cape Floral Kingdom, the only one of the six plant kingdoms that is based in a single country. The Cape Province has the highest known concentration of plant species in the world. The South American rain forest, next in line, has about a third of the number of species. At the time of writing, around seventy percent of the 9,600 plant species identified in the Cape grew nowhere else on Earth. Some of the species lived in such restricted areas that planting a new block of vines could wipe out an entire species.

An old apartment in Cape Town city centre became our home for a few days. The first morning, as I washed the red dust of Kwazulu Natal off my running shoes, tears filled my eyes. I could wash the dust off my shoes, but I would never shake it off my feet or my heart. I missed my first home.

I sat down near the front door to put on my freshly cleaned runners. Sean joined me, looking agitated.

'How could you prioritise meeting a journalist over going to the aquarium as a family?' he demanded as he dropped his bag to the floor with a thump.

'How can you question me about this?' I replied. 'It's an opportunity to promote our tourism and my books. We can't reject free media.'

Sophia and Ellie joined us. Sophia gave me an unhappy look as she put her shoes on. Ellie followed with an even darker look. Since she was tiny, her look could slay. They were upset that I wouldn't be with them. I felt torn between work and family, something all working parents can relate to.

'I thought we had agreed that the aquarium visit was a one for us to do as a family,' said Sean, not hiding his disappointment.

'It was. But then I got this journalist request, and this is the only time we can meet. I want to come with you to the aquarium, but I can't be in two places at once. We'll walk down together, and we'll meet for lunch afterwards,' I said.

'But this is a family holiday. It's important to do things like this together,' said Sean.

I hadn't thought about saying no when I got the request. Now I wondered if I was putting my priorities in the right order. A half-hour later, as I sipped my cappuccino at a chic café in the Victoria and Albert waterfront complex, I tried to seem normal and positive for the interview, but I felt shaken by my family's response. When the interview was over, we met for lunch, and I could still sense their disappointment.

I was upset that Sean made me feel bad for going the extra mile for our business, but also wondered if I needed to take on the message in 'The Hobbit', that gold was nothing compared to loved ones, myself. Perhaps he was right. I needed to think about what I was prioritising in my life. I argued to myself that if I had not focused on growing our business, this trip would not have been possible. Seeking balance was not a simple task. I tended to extremes. Having Sean pull me back was essential, but I resented him for it.

We had great turnouts for the Cape Town events at Kalk Bay books, Claremont Exclusive Books, and Wordsworth Books in Noordhoek. All three stores had made author events part of their fight against the online tide, and it showed. I loved talking to readers.

After the events we changed gear from books to wine, with an overnight trip to Franschhoek, a famous wine region, an hour from Cape Town. It was settled by 300 French refugee Hugenots in 1688, hence the name which translates as 'French Corner'. Late afternoon Sean, Sophia, and Ellie basked by the pool of the B&B while I went for a run. Modest homes pushed up against luxury houses that were gated and alarmed like Alcatraz, their walls bristling with metal teeth and 'armed response' signs. Within a kilometre of the centre of town, tin shacks appeared making the modest houses look like mansions, and the mansions look ostentatious. I turned towards Franschhoek centre following what I hoped would be a loop back to the B&B.

The street was flanked by houses on one side and a wooded park on the other. The memory of being attacked alongside a similar park in Johannesburg twenty-five years before flashed irrationally through me. Run. Run faster. Back then a man dodged in front of me, and I leapt away but he was too quick. He grabbed my arms and dragged me into the park. My body flooded with adrenalin and cold fear as I screamed and struggled. I ripped an arm free and pulled away, but he caught me again. He pulled me deeper into the shaded park. Cars flew past, people could see what was happening, but didn't stop. The hope of help decreased with every step away from the busy road.

Then a car pulled over. The driver leapt out, raised a pistol above his head, and ran towards us screaming expletives and shooting into the air. My attacker let go and fled. I wobbled towards my saviour rubbing my arms, eyes wide, repeating 'thank you', over and over.

'Do you live nearby?' he asked.

I nodded.

'I'll take you home.'

'Don't worry. I'll be fine,' I said.

'No, I'll take you home,' he insisted.

'Thank you,' I said, following him gratefully.

The Toyota held his wife and four young kids, bright faces with eyes wide, after witnessing their dad's heroic rescue. They shifted over so I could squeeze into the back seat. Not long before, South Africa would not have allowed me, a white woman, to travel in a bus with these brave brown people who had saved me. What an insane society we had grown up in.

'Thank you so much,' I said.

'Oh, it was nothing,' he replied.

'Do you always carry a gun?' I asked.

'I keep it under my car seat. You never know.'

I gave him directions to my apartment about two kilometres away. We pulled up to the security gates.

'Are you sure you'll be okay?' he asked.

'Oh yes, absolutely fine,' I said, believing it.

'We'll wait until you're safely inside.'

'Thank you. Thank you so much.'

I closed the gate and waved, then walked across to the apartment and let myself in. The minute the door was locked the shock hit. I walked back and forth as tears streamed out of me. After an hour of circling the apartment, I pulled myself together enough to call my sister Foo. I was not okay.

Now the mere memory pumped adrenalin into my system, and I raced away from the park. Soon houses surrounded me and I felt safer. I looked back and saw the trees of the park in the distance. There was no danger there, just sharp shards of memory.

I was starkly aware of how extreme the difference between wealth and poverty was. Shanty towns lined the highway as we drove from Cape Town airport towards the city. Official signs warned cars not to stop for anything. There had been many hijackings on that stretch of road. Apartheid had left a terrible legacy including extreme inequality, that results in many horrors, including more violent crime[7].

In South Africa, the Gini coefficient, a measure of income distribution, was one of the highest in the world. The higher the Gini coefficient, the greater the gap between the incomes of a country's richest and poorest people. At the time of writing, the Gini coefficient in

South Africa was around 0.63, in France it was 0.32 and in Ireland around 0.30[8].

I found forgiveness from my family in our days in Kalk Bay. We hiked the mountain behind the village, swam in tidal pools, and brunched at Kalk Bay Expresso, greeted joyfully by the people that ran it with humour and care. The sea brought a sense of lightness. Sean and I held hands as we walked to Fishhoek. I was reminded of our carefree early days together, of how we loved each other.

Our visit to South Africa was rich in connections and culture, vibrant, and warm. Back in cold, wet France we were thrown into work with the annual bottling and tourist season preparation. Sean and I slipped back into our work partnership forgetting that we needed to invest in our love relationship, not unusual for couples in the phase of life we were in. We focused on work and children. We ignored each other. It was a dangerous passage that some didn't make it through.

CHAPTER 4

Tubing in the time of growth

S pring growth and visitors quickly brought us into a work rhythm. Sean's days passed deep in the vineyard, mine in the tasting room. Before dawn, I was out in the office replying to emails and bookings and setting up the tasting room for the day. After the visitors left, I worked late cleaning the tasting room, running the dishwasher, repacking, and starting all over again. In his spare time, Sean expanded the vegetable garden. He planted more permanent fruit trees, pears, peaches, apples, and quince, between the beds.

Ad, our friend that had witnessed the attempted seduction of Sébastien, and his wife Lijda, arrived for a spring visit. As I greeted them, I sensed a strange coldness between them. They went down to park their camper van in its usual spot, near the potager, and I wondered what was going on. A while later, I walked down to the clothesline that backed onto the potager with the washing basket. Lijda appeared. She had a knack for joining me when there was something to do, to give a hand, whether it was weeding the garden, harvesting something, or, on that particular day, hanging washing.

Doing a chore together offered space to talk, to share intimacies. Halfway through the basket, she shared the reason for the hint of coldness I had sensed.

'Ad forgot my birthday the day we drove down. Then he did nothing to fix the situation. Of course, it's a lot more than that, but that was the final straw,' she said.

'I'm so sorry Lijda,' I said, feeling her pain.

'He forgot my birthday completely. How do you think that makes me feel?' she asked.

'Not good,' I replied, reflecting both how she felt, and Ad's mistake.

'Exactly.'

'Have you spoken to Ad about it?'

'He knows I'm upset but I don't think he knows what to do about it since he has left it so long. Anyway, I've booked a flight back to Holland this afternoon. We need space.'

I was shocked that their relationship could be shaken. Lijda didn't say it, but I knew she was also worried about Ad's health. She wanted him to watch what he ate and keep fit. Sometimes he ignored her. He was one of the fittest retirees I knew, but Lijda was fitter. She had beaten cancer years before and took health, and what you could do to stay healthy, through diet and exercise, seriously. She knew first-hand. But it wasn't about any of these details on their own. It was about taking each other for granted. Something that happened with time and familiarity. Something that could unravel a long relationship.

Lijda left and Ad wandered around in a fog. He wasn't sure how to fix what had happened. He loved Lijda and missed her, but he didn't know what to do. I suggested writing a letter, a real letter, not a text or email. That evening Ad wrote to Lijda and the next morning he went up to Saussignac to post it. Lijda received his letter, and they started talking, rebuilding their relationship. A few days later Ad left for the long drive back to Holland.

Ad and Lijda had built a life together. They met in Cameroon in their twenties, working on development projects, Lijda as a nurse and Ad as a mechanic. They had four wonderful children and a growing tribe of grandchildren. They had proven wise and loving advisors to us over the years. I sometimes felt like Lijda was a soothsayer, that she suggested solutions before I even saw the problem.

I felt bewildered at the rift in their relationship. I thought that once you had a few decades of partnership behind you, you were home-free, and marriage was a cakewalk. The reality is that a relationship is a constant construction with no room for complacency. Relationships are complex. They need care. A long time together is no certainty of a future together, as I could see in many examples around us.

North American friends invited us to join them at their *gîte*, a self-catering holiday house, in Saint Cirq Lapopie in the Lot department, southeast of our farm, which was in the Dordogne. Sean volunteered to stay to look after the animals, our beloved dog Dora, the cats Guims and Leah, and the chickens. Then Ellie said she would stay too, so Sophia and I set off on a mother daughter adventure.

En route, at the village of '*Faux*' , the word for false, we giggled at the name. At the end of the village, the name sign had a red cross through it, as was custom in France. We stopped for a photo. '*Pas faux*' was a way to say, 'you're right' or 'I see your point'. Then we passed through a village called '*Espère*', meaning hope, and again we laughed at the sign crossed out, 'no hope', but we didn't stop. We didn't need that kind of negativity.

A few kilometres past Espère, a car pulled out in front of us, forcing me to hit the brakes. After swearing at their dodgy driving, I drew back to allow a safe following space. In the distance a stag loped across the field. It felt like a scene from a movie; I had never seen one in real life. As we took in his rippling muscular body, majestic antlers, and sense of otherworldliness, I realised the car in front of us hadn't seen him. It was too late to do anything. The mighty stag leapt at the last second clearing the bonnet by millimetres. He galloped off unscathed. The car pulled over and we saw the driver crossing themselves as we passed. In the rear-view mirror I saw them pull back out in the direction they had come. Everything was okay. It felt like we had witnessed an act of magic. We continued onwards, our hearts racing and our spirits singing hallelujah.

The *gîte* was a walk from the river with a large balcony looking over fields to the cliffs where medieval St Cirq perched 100 metres up. Our friends had eaten an early lunch but offered us something to eat. As we ate cheese, bread and salad, Jenny chatted with Sophia and I. Francois, sat on the terrace reading, and their daughter Amy, disappeared upstairs. I sensed a tension in the air, like we weren't as welcome as the invitation had seemed. Perhaps we had made a mistake coming. As I washed our dishes and thanked Jenny for the lunch, Francois came inside.

'What are we going to do this afternoon?' he asked.

'I'm easy,' I said. 'We're in your hands.'

'How about going up to the village?' suggested Jenny.

'It's too hot for that,' replied Francois. 'How about we take the tubes and float down the river then walk back?'

'I don't want to do that,' said Jenny. 'And there aren't enough tubes. But you go. I'll stay here with Amy. I know she won't want to go either.'

Before we could explore more options, Francois had the tubes, and we were off. Dressed in sandals, swimming togs, light dresses, and hats, we followed him in the direction of the river. As soon as we stepped away from the house, I felt the tension dissipate.

I hadn't tubed since high school and Sophia had never tubed. Francois was a regular 'tuber' but he hadn't been on this river. We were going in blind. At the water, Francois launched himself in without hesitation, while Sophia and I edged in.

'Throw yourself into the centre,' he shouted, bobbing downstream, then swirling joyfully.

I launched. My bum hit the ground, bounced up, and I took off into deeper water. Not wanting to leave Sophia behind, I reached down to grab a rock anchor. The current dislodged my grip as she threw herself in. She caught up. Ahead of us, Francois dug a beer from his pocket, opened it, and lay back to enjoy the scenery.

Crystal-clear water flowed over rocks interspersed with long riverweed that swayed like green hair in its languid passage. A midnight blue dragonfly flew past then hovered nearby, followed by a shimmery lime green one, then a bright turquoise. It was like watching the dance of the sugar plum fairy. They swirled and hovered, tippy-toed on the water, then flew offstage. A white stork landed on a fallen tree and watched us.

Round the next bend a cliff towered to our left with the village perched on top, the 16th century church of Saint Cirq Lapopie floating at the highest point. It was one of the hundred most beautiful villages in France and had been voted the favourite of them all a few years before. To reinforce the glory, the cliffs and village reflected on the water below, offering a double view. A series of small rapids interrupted our rapture. We tumbled through, splashed and scraped, happy to find ourselves on another expanse of slow water. I lay back and looked up at the sky enjoying the sensation of doing nothing in a natural wonderland.

A boat loaded with tourists motored past, diesel fumes in its wake. They waved. We waved back. The stork that had been following us took off, disturbed by the noise. We entered a stagnant zone where we had to paddle to move. The cliff wall hugged the river on the left and dense forest on the right. I wondered how far we had come and how we would get out.

Francois disappeared, taking the left lane around an island in the river. We followed. It turned into a canal with steep sides. I called to Sophia to speed up so we could catch him up. We had no phones, and I didn't have a clue how to get back to the house. My sense of direction was dodgy.

Seeing a steep drop ahead I decided we had to exit. We clambered onto the small edge of the canal. The only way out was up a three-metre muddy wall. Leaving Sophia with the tubes I scrambled up using branches and whatever I could grip onto. At the top, relieved to see a path through the trees, I took the tubes from Sophia then reached a hand down to help her up.

As we turned, a voice from the towpath said, 'What ya doin'?'

It was Francois. He had escaped further on and was looking for us.

'You need to see the sculpture on the cliff wall a bit further on at Bouziès. You'll love it,' he said.

In the 1980s, a sculptor from Toulouse, Daniel Monnier, created a bas-relief in the cliff that hugs the river. His initial authorisation was for two square metres. Today the sculpture covers forty square metres, an escapade of abstract swirls, waves, and creatures, that reminded me of ancient art at Newgrange in Ireland and cave art of the Dordogne. It filled my soul and my senses. I could have stayed observing it for hours, but it was late, and we had five kilometres to walk.

Halfway home we drank rose cordial and carrot juice at a rustic cafe run by hippies. A dusty wood cabin with an armchair converted into a dry toilet served as the 'WC'. It was so eccentric it seemed unreal. As we left, I noticed the upstairs balcony was strung with hammocks for sleeping outside. They were living light on the earth and close to the stars.

Near the end of the forested walk, a mighty tree, deep shocks of bright green moss covering its bark, towered over the others. I reached out to touch an exposed section of rough bark and felt energy pulse through me. I encouraged Sophia to do the same and she looked at me like the crazy mother I was. It was the perfect end to an enchanted afternoon. Soaked in bliss we crossed the bridge to home.

I poured Feely La Source red wine for Jenny, Francois, and me, and we settled on the balcony to enjoy the changing light. As the evening progressed, aided by the luscious Merlot Cabernet Sauvignon blend, the ice thawed. Francois barbequed sausages with courgettes from our garden. We talked about vegetarianism. Francois was into yoga and meditation, but they were not vegetarian. Jenny was an athletics champion and ran at least ten kilometres every day, often more. Amy didn't like courgettes, so she had her sausage on a baguette. Sophia was not a big fan of courgette either, but she ate it. After cheese, Francois suggested we each read something we had written, or read, anything, even a single sentence of a haiku.

'I wanna go first!' said Amy.

'Okay,' said Jenny. 'You go first.'

'Save the plants,' said Amy.

We all laughed. I felt the tension evaporate in the well-being of the night. Francois read an excerpt from a novel he was writing.

'Age is a number given to us by society,' said Jenny.

I read a paragraph about going to the market from my book in draft at the time. Sophia read a favourite haiku.

The next morning Jenny and I went for a run. We chatted about her training, her work, Amy's school. I couldn't help feeling there was an elephant in the room. When I was ready to turn back, she took off to do another five kilometres. I missed having a girlfriend nearby to run with. At the house, I showed Francois the Five Tibetan Rites, exercises I did every morning and that I had become addicted to after my brother-in-law introduced them to me, a couple of years before. Later, at the river, the swim zone, the campsite, and camping parking were packed with people. It was a different Saint Cirq Lapopie from the one we had experienced the previous day.

Sophia and I packed up and headed home. We had been away for 36 hours but back in Saussignac, I had two hours of emails to catch up, pressure for a VAT return, and a year-end stock count to do. I felt like shouting 'STOP'. Pressure rose through my body, but I put my head down and plunged into it.

That night light stomach cramps started. By the next day, they felt as bad as birthing labour. Fever shook my body. Was it a bug I had picked up tubing? Sean tried to comfort me but there was nothing he could do. I took paracetamol and cancelled the tours booked for that morning. The following day I got up and took the tours but still couldn't face eating. That night I ate plain rice. Being sick was no fun. The bout reminded me that we didn't have resilience. Like a monoculture farm, there was no backup if I was down. The season was flying ahead, the grapes were ripening on the vines. Soon it would be harvest, one of the most intense times of year for us. I needed to be well and strong.

CHAPTER 5

Gastronomy in Paris and Périgord

For ten years we had barely left the farm. Our trip to South Africa had been a stand-out and it felt good. At Château Feely we were always sucked into something. Sean saw his vines and his garden from the window and couldn't turn a blind eye. Tasters dropped in; guests knocked on the door with questions.

For our health and well-being, we knew we had to be more organised about taking breaks. In late August, a friend house-sat for us, and we escaped. On the Atlantic coast near St Jean de Luz, armed police walked the sand beside us. An attack on tourists on a Moroccan beach, and a spate of terrorist attacks in France, meant beaches were on high alert. It felt strange. I walked a half day of the coastal path north and then south of our campsite. The sea air and the views of the wild Atlantic refreshed my senses and my soul. I wished Sean would walk with me, but he didn't feel comfortable leaving our daughters alone.

We returned rejuvenated. The new school year and harvest were waiting for us. Sophia and Ellie were on the same bus for the first time. When we walked up, they both had their earphones in listening to music and when we walked back, they talked together, excluding me. Sophia started basketball on Fridays. Ellie said she preferred to walk home on her own those days. Then it extended from Friday to all the days. They didn't need or want me anymore; they were growing up. I stopped walking with them with a twinge of melancholy. The walks to school were precious memories from their primary years. But there was no time to dwell on my feelings. My daughters passed into adolescence and I into harvest.

Every year we faced the inevitable uncertainty of working with nature. Sean and I debated whether to do a sparkling rosé. Our oenologist thought we would not get the sugar, acidity, and flavour to match up for sparkling wine from Merlot grapes. The analysis came back perfect. Perhaps the conditions of the vintage were the reason, and we would not see them again, or perhaps the north-facing slope was the reason. Either way, it was exceptional, and we grabbed the opportunity to add this new product to our range.

'Quick. Get in touch with the harvest team,' said Sean.

Roddy and Cathy, staying in the Wine Cottage, one of Feely farm's self-catering guest houses, volunteered to help. A Romanian couple who had worked with us for a couple of years sent their parents, hard workers with ready smiles, Gérard, a keen hiker and bon vivant retiree, and

Vincent, a gentle, good-humoured local, completed the team.

Picking was convivial. Chatter and song broke out intermittently. Sean and I carried bins up to the tractor together and I felt a connection with him I hadn't felt in a while. Hand harvesting created that vibe. Between the sparkling rosé and the white grapes, I took the TGV, short for *Train à grande Vitesse*, France's Intercity high speed rail service, to Paris for two days to see my sister Foo. The train was packed with commuters but silent. As I navigated the Paris subway from Montparnasse to Porte de Versailles, I marvelled at Metro names that evoked thoughts of royalty and revolution.

Foo glided across the hotel lobby like a catwalk model. We embraced, our atoms colliding like magnets. She was desperately chic in a cream shirt, high wasted cream and black striped skirt and Yves St Laurent stilettos that would have left me flailing. I checked in at reception, and we made our way up to the rooms talking non-stop, about harvest, her seminar, my trip.

'You should have taken a taxi!' Foo admonished.

'The Metro is faster and cheaper,' I replied. 'I got my ticket on the TGV before arriving in Montparnasse, it was superfast.'

'Still Dalling!' Foo often called me 'Dalling', her special slang of darling, one of many nicknames. 'You should have taken a cab. I would have paid.'

Foo had already picked up the tab for my hotel room. I would have been happy to share a room. We were so different and yet so close.

'One of my favourite memories of Paris is catching the first Metro of the day from Concorde to

Montparnasse. Concorde is near Angelina. Remember the hot chocolate?'

'Oh my God Leroy! We're going there tomorrow,' said Foo. To Foo I was 'Leroy'. A long story. We're big on nicknames as I said.

'Amazing! I can't wait to have the chocolate again. Anyway, that day, it was so early, that Angelina and the other shops were closed. The street was empty, but down below, the train quay was filled with commuters. It felt like a party. Greetings rang out. They were workers of the city, cleaners, night guards. People who caught the five-thirty every day. They knew each other. I was blown away by the camaraderie. It was such a contrast to the silence of office commuters.'

'Hmm,' said Foo, not convinced of the advantages of the Metro, or any public transport. She disappeared into the bathroom to touch up her make-up. Her phone rang.

'Answer it, won't you Leroy?'

I picked up.

'Do you accept the charges for a call from Prison X?'

I passed the phone to Foo. She accepted the charges and said hello, then introduced me and handed back the phone. It was my first time talking to my brother-in-law. He was in prison and had been for years, since soon after they got married. While allegedly driving over the speed limit he was involved in an accident where someone in the other car had died. Years later, his court case was still in limbo. He could have been so bitter, but he sounded upbeat.

I felt a deep sadness, for him, and for Foo. Foo had not seen him in person in years. When she visited him in prison, she had to speak to him over a video system from a different building to the one he was in. It was hard to believe the conditions.

We dined at a small, unmemorable Italian restaurant a short walk from the hotel. There was no organic wine, we were both tired and a pallor of gloom hung over us after the call. In the morning, I bought coffee in the lobby and took it to Foo's room. My black jeans, organic cotton long-sleeve top, and leather hand-me-down jacket did not take long to get into. With a quick trace of eyebrow pencil and a swish of lipstick, I was ready. Foo took longer and it showed. She was beautiful without make-up, but in her torn designer jeans, bright pop art top, black leather jacket, and delicate, but expert, make-up, she looked stunning.

Despite my Metro mantra, Foo was not risking it. A taxi dropped us at Angelina tea house. Leather chairs sank into thick carpets surrounded by richly decorated murals and moulded filigree ceilings studded with chandeliers. It oozed early 20th century Paris opulence. I ordered muesli and fresh cheese, alongside their world-famous hot chocolate, from a menu with prices to match the surroundings.

Foo's friends, energetic entrepreneurs, joined us. Like Foo, they were in Paris for a seminar with their business coach. As I listened to their conversation about the conference, I found my mind wandering to the waitresses. Dressed in French maid outfits, black with white trims, they seemed stressed and rushed. One arrived with a tray loaded for us and I thanked her profusely. From working as a waitress when I was a student, I had an appreciation for the demands of the job. The hot chocolate was so thick it poured like a lazy river of mud from the jug.

I used my finger to scrape the last vestiges of chocolate and caught a look from Foo. I smiled guiltily and picked up my spoon to finish catching the last few drops. Then it was me and Foo, exploring Paris together. We visited Shakespeare and Co, an English bookstore started in 1951 by George Whitman and a haven for writers and readers. I could have stayed there for hours, but pangs of hunger drove us onwards. For all its richness, the food at Angelina had not filled us. I found an organic vegan restaurant that looked perfect for a late light lunch. Foo was reticent, she did not expect good things from a tiny place promising vegan food. But the chickpea mousse and salad alongside a fresh apple, carrot and ginger juice for Foo, and the falafel and salad with carrot, beetroot, and parsley juice for me, were bursting with flavour and goodness. We felt happiness in our bellies.

Sean texted 'We must pick Monday, get the team lined up.'

At that moment in our farm cycle, every hour mattered. Back at the hotel, I organised the team via text and email then picked up a conversation with the concierge as I waited for Foo in the lobby. He was impressed that I was a *vigneronne*, a winegrower. Our *métier*, job, was a source of national pride, something most French people identified with. Chatting with people in Paris on that trip, the concierge and our taxi drivers, I felt like I was part of France, part of continental Europe. It felt like home for me, the same as I felt about Ireland and South Africa. It was a welcome sensation.

The gastro-bistro restaurant Foo had chosen for the evening was dark and cosy. Small tables were crammed in next to each other and dark red velour covered the booth seats and chairs. A brass rail next to the door held wooden coat hooks. I felt like I had stepped into Diagon Alley, the shopping street in Harry Potter, and the menu packed with traditional Burgundy fare of frogs' legs and snails served to confirm it.

An *amuse bouche* – an amusement for the mouth – of cucumber salad and tiny cheese puffs whetted our appetites. I perused the wine list. Seeing none marked organic, I asked the sommelier which wines were.

'We only have one organic wine,' he said. 'I don't like to have organic wines since one out of two bottles are bad. I can't take a risk like that with my wine list. But half of the list is biodynamic.'

'Oh,' I reacted.

'Yes,' he continued. 'A lot of the biodynamic growers don't want to follow the organic rules and be certified, they don't want to risk their crop.'

'I'm an organic and biodynamic winegrower here in France,' I said with a wry smile. 'A grower must be certified organic before they can become certified biodynamic. If half are biodynamic then at least half are organic. In biodynamic we follow even stricter rules than organic.'

He looked surprised that someone that he thought was a foreigner without a clue, knew something about organic and biodynamic rules in France. He left us to find the white wine he thought was organic. It was, and delicious too.

Our starter salads arrived, tomato and goats' cheese for me and langoustine for Foo. We slurped them up, savouring the conversation, food, and wine. At the interval between courses, I requested the *carte des vins* again. In the selection of affordable wines by the glass was a Bourgueil, the village where we stayed on the first trip to France when we met Deboussey. I caught our sommelier's eye and asked for a glass 'if it's organic'.

'I'll check,' he said. He returned with a delighted smile and showed me the organic EU leaf logo, guarantee of organic practices in the winegrowing and the winemaking, then poured. I leaned into the glass and was transported back to our holiday. I passed it to Foo, who took a sniff, then a sip.

'Incredible. It's like we're back in Bourgueil,' she said. 'My martinis don't do that.'

Foo drank martinis rather than wine since wine gave her a hangover, a lament I often heard. I always asked; 'was it organic?' The pesticide residues in some chemically farmed wines were sure to have a bad effect.

We luxuriated in the wine, in its ability to take us to places with a simple sniff. By the time we finished the cheese course, we were talking about how to create online courses and how I could digitise some of the classes we offered. I made mental notes and followed up with physical notes the following morning as we chatted over coffee. Foo had developed her online business, offering virtual coaching and mentoring, from nothing to a full-time career over the previous couple of years. She assured me if she could do it, I, who had a successful career in technology consulting prior to becoming a winegrower, could definitely do it. She planted a seed that would save our livelihood a couple of years later.

The next day at the *Musée D'Orsay* Impressionist Art Museum, we strolled together then separated after agreeing a time to meet. I wandered through a field of sculptures then climbed the back staircase where a quote stopped me in my tracks. I recall it as something like, 'Harmony in nature is when the means to get there is in perfect balance with the result.'

Seeking harmony from what nature gave us was exactly what we did as winegrowers. We sought equilibrium in the vintage by adjusting our work, and the wine we made, to the weather and the conditions. Winemaking is art and science. It is nature and emotion. Our quest, as natural winemakers, is to create wines in balance with the season not to a fixed recipe. I continued climbing the stairs then circled back to the entrance to meet Foo. The doors were manned by heavily armed guards not the usual ticket checkers. Reinforced security reminded us of recent terrorist attacks.

Foo and I wept as we hugged goodbye in the station. Leaving was hard. We missed each other. Tissues in hand we headed to our respective destinies on opposite sides of the earth, her in Los Angeles, me in Saussignac.

Back on the farm, harvest grabbed my attention. At dawn, the grapes were so cold my hands froze. The sun popped over the horizon, its glow promising another glorious day. We had energy in the morning, the team moving up and down the rows, like a flock of birds, separate but together. By midday, the sun was baking, and we moved slowly, like the stickiness on our hands was sticking us to the ground.

There was community in singing together, the feeling of the year coming together in a joyful crescendo. The quality was better if we picked by hand, but it was more than that. The previous year, in a couple of hours of picking opposite Cécile, our apprentice at the time, on these rows of ancient

Semillon, I learnt more about her hopes and dreams, more about who she was, than in months of working together.

At the end of harvest, we took the team to a local restaurant to celebrate. In Sean's thank you he reminded us that it wasn't the weather we remembered most about a harvest, it was the people. Each wine brought back memories of those who had helped make it a reality.

Friends from Ireland, who had picked with us years before, dropped by and invited us to celebrate the end of harvest with them at a nearby Michelin-starred restaurant, *Les Fresques*, at Château des Vigiers. The symphony of food melted our harvest exhaustion away. Oysters with granny smith apple and Neuvic caviar were paired with Feely *Sincérité* Sauvignon Blanc. Neuvic caviar is produced in the Dordogne on 20 hectares bordered by two rivers, the *Isle*, and the *Vern*. The two rivers provide quality fresh water and the fish are fed on organic food and stocked at low density. It is an unusual Dordogne gourmet success story to complement the more traditional gastronomic riches – duck, strawberries, truffles, walnuts and wine – of the region. The minerality of the *Sincérité* from the limestone made of ancient, compressed seabed, and its acidity, offset the oysters and caviar perfectly.

The sensations continued with poached egg with walnut and truffle and Feely *Générosité* barrel-aged Semillon, a rich white wine. Walnuts can be pickled green long before maturity or picked at maturity, and crushed for their oil or used simply as nuts straight up or in a recipe. I adore a drizzle of walnut oil over fresh figs on a slice of local bread. Cracking homegrown nuts and then studding them into local honey spread thick on a chunk of homemade bread brings satisfaction to hands, eyes, and stomach. At *Les Fresques* we were experiencing the local produce transformed into elegant dishes, but the wonder of the gastronomy of the Périgord – our region's traditional name – is often its simplicity. Sublime fresh local products with no need for artifice.

Truffles are called the black gold of the Périgord due to their value. In 2022, they were selling for around €1000 a kilogram. The aromas and taste are like something between heaven and earth, both floral and earthy. A small quantity of good quality fresh truffle can transform a humble dish into something extraordinary. Dust a pumpkin soup, a risotto, or fresh buttered tagliatelle with truffle shavings, and you transform it into a gastronomic delight. In my book, 'Vineyard Confessions', we stay on a truffle farm near Cahors and explore this exceptional product of the Périgord.

The *Fresques* voyage continued with melt-in-the-mouth fillet steak, grape and beetroot sauce, and Feely Grace, a blend of Merlot and Cabernet Sauvignon. The fillet was salty, buttery, and flavourful, the beetroot earthy, and the touch of grape pulled it together with a hint of sweetness.

Fresh strawberries and intense chocolate ice cream were followed by coffee and petit fours, a fig macaroon, a tiny pear tart, and a chocolate truffle. I had a feeling of well-being from the luscious food and seeing our friends, but Sean and I were struggling. We thanked our friends for the luxurious treat and made our way home in silence. We had been working 'hand in hand' over harvest as we had to, but we never held hands. Outside of work, we had nothing to talk about. The moment of carrying the buckets together and feeling collaboration at the start of harvest had been an exception. I felt a slight sickness in my belly, as I reflected on it on our way home. Seeing our friends so happy in their relationship brought the problems in ours into stark relief.

During the week we rarely ate lunch at the same time as I was racing with groups in the tasting room and Sean didn't like to be rushed. At night we took turns cooking and then ate with our daughters but concentrated on them and their lives. I went to bed early and Sean stayed up late. We hadn't been on a date night in years. There was no physical contact. Sean grunted hello at me. We made little eye contact. We were losing touch with each other.

As the days shortened, harvesting passed the baton to vineyard maintenance. COP22 slipped by, hosted in Marrakech, Morocco. There was limited media coverage compared to COP21 in Paris. Fossil fuel lobby groups were given places with observer status and walked the halls alongside climate activists. The November average carbon dioxide parts per million (ppm) in the atmosphere, recorded by the NOAA[9] at Mauna Loa Observatory in Hawaii, increased from 400.27 in 2015, the time of COP21, to 403.72 in 2016, the time of COP22. The year of COP21, 2015, was the first time the November average ppm was over 400 since the recordings started in 1958. Research suggested that the last time the concentration of carbon dioxide was averaging this level, was more than two million years before, when horses and camels lived in the Arctic, and seas were more than nine meters higher than today.

Seas were rising slowly, in centimetres rather than metres, but reading the nine meters gave me a shock. My parents lived on the coast of Vancouver Island in Canada. I looked up the elevation of their house and felt relieved to see it was well above sea level. They laughed when I told them about it, and suggested we meet up before we all disappeared under the sea. They were planning what they thought would be their last trip to South Africa. On the cusp of 80, but healthy, they were plotting two months to see friends and family flung across the country. We decided it was an unmissable opportunity to see Sean's dad and my parents on one trip. We hadn't seen my parents in years. I consoled my cognitive dissonance about the carbon dioxide emissions with the fact that we had planted more than a hundred oak trees.

The trip would be on a tight budget, with no fancy tents. I sketched out our voyage, the first week with Steph, then Sean's family, and the second week with my parents in the Eastern Cape, renting a friend's rustic beach house. I hadn't found a new apprentice, but I put that worry aside. Francine Klur had invited me to attend the International Biodynamic Women's Conference in Alsace. With that trip to Northeast France and our holiday in the pipeline, we had a reason to speed up the winter pruning.

On the voyage to South Africa, I would reconnect to a physical place that was part of me. In Alsace, I would question some of the foundations of my life and my place in this world. I would realise that I needed to work on developing my sense of place in the 'everywhen', what the Aboriginals of Australia called being in meditation and sensing your place as part of the universe. That discovery was coming, the next phase was written in the stars.

CHAPTER 6

Confronting myself in Alsace

The trees were dropping the last of their leaves as I packed my suitcase for the conference in Alsace. Sean chauffeured me to the local station, completely deserted at 5.30 a.m., unlike Concorde in Paris. He got out of the car to check I had everything. We hugged. Leaving was a reason to do something we rarely did. He left me to cross the track to the silent, empty departure quay alone.

A little later, as the TGV gained pace out of Bordeaux, I began reading a Steiner text recommended as preparation for the conference. Rudolf Steiner was the father of biodynamics. Many people had heard of biodynamics through wine, but it was valid for all farming, and went well beyond farming, into philosophy, architecture, health, education, and spirituality. Steiner recommended no consumption of alcohol. In an intriguing twist that he may not have liked, wine was the product that had raised awareness of biodynamics in modern times.

The text was one of Steiner's spiritual lectures. It had nothing to do with farming. It was strange. Too strange. I put it aside and reached for my breakfast, a hunk of homemade bread and a few squares of dark chocolate. The purity of Gilles' organic flour and the smooth dark cocoa blended in my mouth bringing a sense of well-being in my belly. It was a healthy version of '*pain au chocolat*' or '*chocolatine*' as the luscious chocolate pastry is known locally. I caught the crumbs on my lap and dropped them into the waste basket. Then I opened my copy of Steiner's Agriculture Course that I had brought to reread.

The Agriculture Course forms the backbone of biodynamics in agriculture and gardening. It was originally delivered as a series of lectures in 1924, created based on observation, research, and interrogation of ancient methods, and peasant farming. In it I found a solid anchor. He outlined the biodynamic farming method in wide brushstrokes, offering ideas, rather than a rigid methodology. Each farm was as different as the farmer.

Steiner predicted that our science of today, the science of looking at something in isolation, rather than in the context of living things, would lead us to bad results. He said this method would make it possible to prove something in a certain context that would be false in another. I could see he was right. To take a simple example, the debate about sugar and fat in nutrition, if we look at them in isolation they look like bad guys, but they are essential to our bodies.

That night, installed in a guest house in Munster, in the Haut-Rhin department of north-eastern France, I opened the conference gift bag and found a small bar of biodynamic cow dung soap. Any aroma was better than mine after twelve hours of travel and four hours of welcome drinks, singing, and dancing. It smelt remarkably good, like the cow dung horn preparation we used in the vineyard, like the forest floor. Showering away the grime of the long journey, I found myself laughing out loud at washing with cow poo.

The next day, as I waited for a lift to the conference, a group of women near me chatted about how much time they needed to get ready.

'I can get ready in less than an hour,' said one.

'Ah, I don't need that much time,' said another. 'I don't shower every day.'

Not showering every day represented a good ecological choice but combined with the dung soap of the night before, it created an eco-warrior cliché that had me chuckling behind my scarf.

The theme of the conference was love and tolerance. What could we do at a social level on our farms and in our lives to advance love and tolerance? The first speaker talked about the terrorist attacks in Paris, Nice and Brussels, in 2015 and 2016. She didn't know what to say to her children about why it was happening, and what she could do to stop it. She felt powerless.

Then a German woman whose daughter had married a Muslim man shared how she had confronted her prejudice and delved into the shared history of Christianity and Islam, and their parallels, which she shared with us. It was interesting but I wondered why I needed to hear this debate. I felt like it didn't concern me. I was not at the conference for religion. I was there for hardcore farming action.

Then a Muslim woman described her experiences growing up in France, how difficult it was to integrate into French society. She explained that when the terrorist attacks happened, like the first speaker, she didn't know how to explain it to her kids, but more importantly, she was worried about how her children would be ostracised at school. She felt that peaceful Muslims, like her family, were seen as part of it, somehow responsible.

Fear took over people's brains and they began to behave in ways they would never have otherwise. I put myself in her shoes and found it did concern me. I began to see the attacks and their consequences in a wider way. She was describing racial and religious discrimination that needed to be addressed. Like apartheid in South Africa, this was something that needed attention. We had to stand up to it, vociferously. We had to be actively anti-racist. It was a fight we all needed to engage in, at our farms, and in our community.

A blonde woman, dressed in traditional peasant clothing, translated the sessions, speaking from French to German and German to French like she was an extension of the speaker. There was a streak of wildness about her. After interpreting for others, she offered a conference session on World War I, with a key message that it was started by the English. I was no geo-political buff; but I knew there were complex issues at the heart of it, far beyond an Austro-Hungarian archduke getting assassinated, and certainly not wholly due to England. When we met later, I felt attacked by her aggressive '*vous êtes anglaise?*' you're English? I did not expect to see this split, this antagonism, between mainland Europeans and the English. Brexit had recently been voted for and felt like it had added to the divide.

I explained I was a melting pot and an Irish citizen. English was my mother tongue, but I also spoke French, and a little Zulu and Afrikaans. She asked me to speak Zulu and went into ecstasy at the click sounds. Throughout the conference, friendly jokes were made about the differences between the Germans and French. Many participants had not travelled beyond the country neighbouring where they lived. They didn't know how to react to me as a French-speaking anglophone. I felt like an outsider, a real *étranger*.

The conference made me starkly aware of how important programmes like Erasmus university exchanges were to fostering understanding of, and between, the diverse cultures in Europe. Despite being offered the chance to remain in Erasmus post-Brexit, the UK had decided to leave the programme. Erasmus is a pan-European university exchange programme to foster links and European integration. Students can spend a couple of months or a full year in another country as part of their degree. With the UK out, EU students cannot attend UK universities on exchange and UK students cannot attend EU universities. It was a loss for both sides.

In the next core session, we were asked to consider what sort of person we were. What was more to the fore; thoughts, feelings, or will/ action? For me, at the time, will and action were to the fore, at the cost of thought and feelings. We were asked to think about this, to think about how we could compensate if we were oriented in one direction. How could we wake up to the needs of others, and include the social aspect in our work and home life, how could we make space for social encounters?

I felt blown open like the conference was forcing me to consider a whole new dimension in my thinking about my life and our farm. It was demanding that I put aside capitalist, economist blinkers, and see the world through a much richer lens, one that included relationships and nature, and allowed more feelings and emotions into my life, and by extension, into our farm and business.

Between sessions, we feasted on delicious organic food. Tea and cake flowed morning and afternoon. For lunch, carrot puree, pan-fried leeks and oat cakes were followed by apple crumble with millet. For dinner smooth parsley soup, black risotto rice and butternut, and cheesecake with a coulis of red fruit, preceded singing and dancing in the big hall.

Despite the fabulous food and the sensation of being on a retreat, I felt disconcerted. Many of the participants had attended the conference for years or even decades. I was a blow-in. The event was started by Maria Thun in the 1950s in response to the Biodynamic Conference only welcoming men at the time. Maria Thun was a leading biodynamic researcher, but despite this, she was excluded. Over time, the conference had morphed into something less about agriculture, and more about biodynamics in a wider sphere. The main sessions were complemented by breakouts on observation, dance, and art. I was expecting farming and business, things I knew about, and I was being served up something completely different.

In these more sensitive areas, I felt like a rabbit in the lights, completely unsure, like my whole reason for being was in question. I began to realise how much I had set aside care for heart and soul in the ten years of establishing our business. But that realisation was too frightening, so I ran from those thoughts rather than facing them. Instead, I promised myself I would never attend a conference like this again. Not for me this soul searching. What was the practical gain? What had it done to advance our farm, our business? I felt torn, like I could see that they were onto something, but was too scared to go there.

Looking back, I can see this was a major psychological moment for me. Our society rated success as financial success and I had been doing the same. Prioritising business over family, what Sean had called me out for, was an example. I needed to be with friends and family, to give myself more heart time. I avoided facing that realisation. I felt exposed. I didn't know how to accept and act on the messages I had received, so I closed them out.

After the conference, as Francine and I travelled back to her village of Katzenthal, I could tell her mind was elsewhere. She had four days of work built up and needed to catch up. Perhaps she could also tell how shaken I was by the conference. Things had changed since we visited *Vignoble Klur* six years before, a visit that was part of the second book in the vineyard series, 'Saving Our Skins'. Clement and Francine had moved from the main property, where the winery, tasting room, and holiday apartments were, to a new house, two plots up the street.

It was a step towards privacy for their family life, one where guests couldn't knock on the door at all times of the day and night. On the gate at the entrance to their new residence was a sign saying: 'please call this mobile number if you desperately need help, don't ring our doorbell'. I knew we needed what they had created. A division between work and home. Sean and I had debated how to do it. So far, we hadn't succeeded in creating this level of privacy for our family life at our farm where we welcomed guests year-round.

The Klur's new dwelling felt like a sculpture. A smooth tree trunk served as the centre of a spiral staircase that took us from the boot room and technical zone for heating and plumbing, to the living space. There, in the compact kitchen, a side slice of an ancient oak formed the island. Beside it, a dining table led to the lounge section of the living space. The wood structure encased picture windows that looked onto a wild area where sheep grazed. They had stringy black and white wool, long winding horns and charming horizontal ears. Watching them grazing peacefully felt therapeutic. Inside, solid wood sliding doors hid cupboards and the entrance to the staircase that kept spiralling up to the main bedrooms and bathroom. Following Steiner's guidelines on architecture, there were no electric points, lights, or technology in the bedrooms, contributing to a sense of peace.

We ate lamb chops and green beans with Clément's mother across the field. *Madame Klur*, in her 80s, made lunch for the Klur team since forever, a way to remain part of the daily action. As we ate, I asked Clément about the sheep.

'They're Racka, a hardy Hungarian breed,' he said.

'How much work do they require?' I asked.

'We do almost nothing. They have an automatically refilled water source. I give them a bit of hay in winter. They are disease-resistant and autonomous. They pretty much look after themselves,' he replied.

It seemed too good to be true. The lamb chops had been raised right here. They made sense ecologically. We had researched having sheep but in our area there was a risk of sheep being killed by roaming dogs. That, and what we thought would be significant additional workload, had put us off. Perhaps we needed to relook at adding sheep to Feely farm.

I noted a quote from the conference that said, 'if we eat food that has been through great suffering, we cannot be well nourished. In a sense we are reduced to materialism, our hearts are destroyed. Then our heads are destroyed by systemic chemicals and what is left is a materialist machine not a thinking feeling person.'

It was imperative to bring kindness and health back to farming. Intensive, industrial animal farming and systemic pesticides hurt us, they hurt our ability to care, our health, and our happiness.

That afternoon, on a solo hike around the '*Trois Epis*', above their house, it was easy to see which vines belonged to the Klurs. They were grassed, not weed-killed, and fenced, so the sheep could graze them. Up there on the foothills of the Vosges mountains, I looked down onto Katzenthal, and around to Ingersheim, like little mini-town villages far below. Walking free in nature nourished my soul. I felt powerful, like some force was filling me with energy, helping me get my equilibrium back.

Francine flew in the door early evening, and we chatted as she prepared lasagne.

'Caro, we got to a point where we had to grow and take on more people, and more pressure, or say hey, that's enough, and follow a new route,' said Francine. 'We decided to take the new route. We didn't want to spend our days on the road, going to wine fairs, making more wine but then also constantly under pressure to sell more wine.'

'A major crossroads,' I said.

'Yes. And we don't know if we've made the right choice yet, but we made it. We sold off most of the vineyards. Our vineyard employee was happy to move to the new owner and continue with the vines. We kept small lots of Pinot Noir and Riesling, and our grand cru. Now we make the kind of wines that we like, natural wines. They are confidential, small lots. We sell out without needing to do professional wine shows.'

'That sounds ideal,' I said.

'Exactly,' said Francine, giving me a nod. We both knew the frenetic round of wine shows to sell each vintage could be fun but was also exhausting, time consuming, and expensive. 'Around that time, my right-hand woman for our tourism activities found her dream job elsewhere. We decided to follow glorious contraction rather than expansion. We are coping with apprentices, volunteers, and short-term contracts now, no full-time employees.'

'What about the accommodation?' I asked. 'It's so difficult to find part time cleaners.'

'Frustrating, yes. I'm thinking of renting the apartments long term instead of as holiday accommodation and creating a workspace, a kind of co-working area. I'm tired of the constant changeovers, of having guests who aren't in our ethos, people who want heating in the stairwell in winter or aircon in summer because they opened the windows and let the hot air in. I want to rent to people who are in the same spirit as us, people who reflect on their impact on the planet.'

'But at the same time, you have done so much to educate people over the years,' I said. 'Welcoming people who are not aware is a way to create change. I remember seeing your notes on eco-living when we stayed in your apartments. When I got home, I did the same. I set up information about the environment, saving water, electricity, composting, as part of our visitor notes. People have had their eyes opened thanks to you.'

Francine looked up from the chopping board.

'You're right. We want to keep making an impact like that. We plan to run courses and offer events with partners. But I don't want to do the delivery, just provide a place for others to do it. I hope that the people that rent and use the co-working space will be eco-businesses and that will create a new dynamic.'

Clément swished open the sliding door, fresh from a painting course, something he could do now that they had shifted gears. We embraced with joy. Over aperitifs of black radish slices dusted with salt and Klur no sulfites added sparkling wine, the conversation was vibrant. We caught up on family, farming, life, and dreams. It was so good to see these friends again.

When Clément dropped me at the station the next morning, I felt sorry they lived so far away. After being with them, I always felt like I could see more options for ways forward in life and work. Leaving Colmar, the view across to the Alps in the east was highlighted by a thread of pink. As the train slowed into Strasbourg, swathes of cherry, fuchsia, and purple burst across the sky. It felt like a symbol for my heart opening with what I had learnt that week. The conference was a wake-up call to think about life in a different way, to reflect on the elements of heart and soul in my life. The realisation and the light show across mountains filled me with transcendent joy. This conference had been a gift at a perfect time, a necessary stepping stone on my journey to finding zen, even though it had created a distinctly un-zen, uncomfortable feeling as I was going through it.

In the station, cold commuters rushed in all directions, intent on their destinations. I picked up a newspaper on the seat next to me. The lead article announced that a terrorist circle in Strasbourg had been exposed the day before. I looked up as a woman with her head covered raced passed, eyes straight ahead, looking scared, like she expected someone to accost her. I felt her anxiety. After the conference, I had a better idea of how she felt. Goosebumps shivered across my body.

A half-hour later, installed on the TGV, a studious-looking, bearded man took the seat beside me and pulled out a copy of a Terry Pratchett novel. It gave me a hook to strike up a conversation with him. He introduced himself and his studies.

'I'm doing my Ph.D. in Strasbourg, researching proteins to create the smell of pineapple. The food industry needs to find new ways to make food smell and taste good now they are being forced to decrease additives like salt and sugar.'

'Perhaps they should use the good, true, original products,' I countered.

'They need to meet constraints of cost and conservation,' said the young man. 'When food has to travel long distances, it needs to be denatured and the flavour needs to be added back.'

His words shocked me. I felt like I had transferred into the parallel universe of one of Pratchett's novels. I would learn more about this murky world of industrial food at the Ballymaloe Litfest a few months later.

My train companion was relatively unaware of organic and what it meant in terms of farming and production. His doctorate was solving a point problem for industrial food. He didn't appear to have thought about what that meant in a wider sense. Despite our different points of view, we chatted for a long time and shared a few laughs. At a station in the Marne Valley north of Paris, he held up his hand to wave goodbye and was gone.

Another young man boarded and took the seat. He looked worried, almost shifty, and smelt of cigarette smoke. I said hello and then carried on reading. He shoved a cigarette packet into the seat holder and then slugged a Red Bull.

I noticed he had two unmarked black bags between his legs. It was a short time since a young Moroccan had attacked a TGV with an assault rifle AK-47, a handgun, and a knife. I felt irrational fear rising in me. This guy looked around twenty years old and could have been of Arab descent. He was acting nervous like I'd expect someone to be if they were about to blow up a train. It was crazy how my mind took off. I was acting out the exact prejudice that I had learnt about at the conference. Even though I knew it, and how wrong it was, I struggled to contain it. I told myself to get a grip, but I still felt the irrational fear. I decided the only way forward was to strike up a conversation with him.

'Are you headed for Bordeaux?' I asked. There were many stops en route, but the final destination was mine.

'Yes, I'm going to take a test to join the army,' he replied. No wonder he was nervous.

'Where are you travelling from?' I asked.

'I come from Reunion. I've been in France for three days staying in a hotel waiting for this test. If I get into the army, it'll be for five years.'

'That's a long time.'

'Yes, but I have to do it, there are no jobs at home.'

I saw tears well in his eyes. He looked away and then reached into one of the packs at his feet.

'My daughters,' he said, showing me passport photos of girls that looked perhaps three and one years old. 'I am not with their mother anymore. I have to make something of my life.'

'Do you know anyone in France?' I asked.

'No, I've only been here once before, when I was a kid,' he said.

'We have a neighbour from Reunion. When you get a chance to get out of barracks, come and visit us. I'll introduce you to Gaby.' I handed him my card, feeling ashamed and embarrassed at the thoughts that had been passing through my head a few minutes before. 'Now you know someone in France.'

We chatted for a while, talking about where he would be in the army and what training he had done for the test. I talked about how we came to be in France. Intermittently between our bouts of conversation I read, and he flicked his mobile phone.

We arrived at Bordeaux station. He got up, ready to go
to his test. He had much to face, but he took the time to
turn to me and say:

'Goodbye and thank you. I will look after your card and
come and see you. Thank you.'

'Good luck with the test,' I said. 'I'll be thinking of you.'

I felt warmth and love for the young man flow through
me. Meeting him was a perfect finale to the lessons about
prejudice, tolerance, and love, that the conference had
explored. I had much to learn.

At home, the pressure was on to finish pruning and
then pulling the wood – taking the cut canes off the
trellis – before our trip to South Africa. We strongarmed
our daughters into helping us. That Sunday was like a
meditation. We sipped tea together, the four of us. Then
Sean gathered us to make hummus from homegrown
chickpeas that Sophia and Ellie had shelled in the summer.
After feasting on homemade bread and hummus, we took
off like a flock of birds into the vineyard to pull wood
together. Sophia and Ellie dragged their feet at the start,
but the sun was shining, and the work was rhythmic. We
moved through the rows like a dance. It was more fun as
a team than alone. We stopped to watch raptors soaring
overhead and marvelled at the new moon in the pale blue
sky.

The next day as I completed the end-of-year accounts,
I felt schizophrenic. We had to grow and take a big leap
of faith with more investment and a full-time employee
or draw back and take the other path like the Klurs had
done. In the state we were in, there was no balance, we were
overworked but not quite at the point where we could
justify a full-time employee.

That winter we didn't have an apprentice, but Nick and Margherita, friends that ran a travel blog called the Crowded Planet, had agreed to housesit for our trip to South Africa. Knowing we had people who loved our animals, and Saussignac, coming to take care of the farm brought a sense of comfort.

In January I was gripped by a sinus and eye infection. I remembered a guest asking whether we ate organic. I replied 'Yes, and since we have followed that I rarely get sick.' That was no longer true. In the previous year, I had been wracked with gastroenteritis and infections. The day we left for South Africa I finished a round of antibiotics, something I hadn't needed for about a decade.

CHAPTER 7

Pilgrimage in the time of hail

S teph and Dave's neighbourhood in central Durban pulsed with big city energy. Despite protective walls and alarms on every property, the park we walked to felt safe. Outdoor gym facilities and paths crisscrossed the recreational area. A young reggae dude taught hip-hop to a group on the north corner.

We dropped onto the grass to watch, soaking up the sensations of this world so far from rural Dordogne.

'I'm crushed by the heat,' said Sophia.

'It's the humidity too,' I replied. 'We're not used to it.'

'That's Durban,' said Dave. 'Humidity and energy. Look at this dude with his hip-hop class. Durban's a buzz. In my agriculture classes, I have people that are more used to the city than to wilderness. I took one of my groups to the Great Karoo for a field trip. The immensity of it, the space, the desert nothingness, the huge starry sky, it scared them. It was too raw, too open. They were used to this city with its energy and noise. They didn't know how to handle the silence.'

'Radical,' I said.

'*Ja*. People here get on with their lives despite the challenges. They make things better, improve their homes. I saw a guy buying a tile at the hardware store. I asked him 'Why only one tile?' He said he buys one every day after work and takes it with him in the taxi, then carries it home. He doesn't have a truck. This way he is slowly getting enough to tile his kitchen.'

In South Africa, a 'taxi' is a term used for the minibus taxis that offer shared transport, like a bus system, to more than half of the country's commuters.

'Economical, and ecological,' I said.

'Exactly. I'm working on a DIY insulation start-up that will make it easy for people to create insulation by mixing water with a base they can buy that is small and light enough to carry home in a taxi. With global warming, we need to insulate houses. It's different to you guys in Europe, where insulation is to stay warm. Here, we need it to stay cool. But most people don't have their own vehicle. Delivering big stuff like insulation is expensive,' said Dave.

'That sounds smart,' I said.

The hip-hop dude wrapped up his class and showed his participants the move he would teach at the following session. We watched as he floated above the ground on one hand then flipped himself back to standing in one fluid impressive motion. I felt like clapping but given it wasn't meant for us I held myself back. With the impromptu show over, we got up and started walking home.

The next morning Steph and I ran the beachfront. We talked non-stop, catching up on the year that had passed and revisiting our plan to celebrate our 50th in Italy. I needed King Shaka Water World where my family was waiting to cool off. Steph left us. Sophia, Ellie, and I found our way to the women's changing room, a big open space lined with benches and hooks to hang your gear while you changed. The room was packed. As soon as a hook came available, I pushed my bag onto the bench below it and we huddled into the space.

Zulu reverberated around us as friends shouted across the room to each other and people wandered around naked. Sophia and Ellie were wide-eyed, overwhelmed by culture shock. Holding towels, we changed. Bathing suits on, we walked into sun so sharp it cut shadows like a knife. We found Sean under a shady tree, an island of calm in the brash brightness.

We slid crazy and free down the second-highest slide in Africa, water spraying into our faces. For hours we did nothing but fun, my mind so far from vineyards that I felt renewed. We tubed the 'lazy river' past tanks of sharks, dolphins, and wild coloured fish. We loved it so much we took it three times. No one wanted to leave. That night we wolfed down calamari and spinach, 'umfino' in Zulu, that tasted of the terroir. Later I would watch the Netflix documentary 'My Octopus Teacher' by Craig Foster and read 'The Soul of an Octopus' by Sy Montgomery and discover a deeper appreciation of cephalopods, the most intelligent known invertebrates, and the family of sea creatures that gave us calamari.

As we drove up to Howick, the weather turned cold, and rain pelted down. It felt more like winter than summer. Climate change wasn't only bringing heat, it was bringing extremes, and it was very unfair. A few years later Durban was hit with devastating floods that killed hundreds of people. According to the 2023 report by the Intergovernmental Panel on Climate Change (IPCC), the United Nations body for assessing the science related to climate change, between 2010 and 2020 human mortality from floods, droughts and storms, was 15 times higher in highly vulnerable regions compared to regions with very low vulnerability[10]. The terrible truth is that often the most vulnerable people have done the least to cause climate change but are paying the highest price.

I met Russ for hot chocolate. He always made time to meet, a friend, generous in every way. We stuck another stake in the ground about celebrating our 50th together. This trip was about family and connections, tending to heart and soul. I knew I needed to take the Alsace conference's messages on board.

Sean's aunt, Sheila, lived on her daughter and son-in-law's farm, a long dirt road drive from the town of Maclear in the Eastern Cape province. Up there, in the foothills of the great Drakensberg mountains, one side of a farm could receive double the annual rainfall of the other and have dramatically different micro-climates due to the mountains' influence. It was raining on every side as we drove in. Sean dodged the puddles but still got stuck in the mud. Sophia, Ellie, and I pushed the car out. Mighty we were. Sean drove off pretending to leave us behind. We laughed.

Sheila welcomed us with warm hugs. I had met her, and Jim, Sean's uncle, when they stayed with us in France ten years before. She moved more slowly now, her eyes had dimmed, but she was a powerful spiritual force. She understood life, its undercurrents, she was a wise one.

Sean's late uncle, Jim Feely, was a renowned conservationist in South Africa. He created the Wilderness Leadership School with, among others, a famous conservationist, Ian Player, who called Jim the 'Brain'. Jim was also an academic, his core research areas were anthropology and ecology. Through studying cave art and archaeology, he showed that humans, specifically the San people, had lived in the Eastern Cape for millennia.

For Jim the spiritual side of wilderness, and conservation of it, were important. Wilderness provided a spiritual place. Wilderness wasn't without humans, it was without 'industrial man'. Jim led walking safaris rather than 4x4 safaris at the Wilderness School. The spirit of the wilderness, not how much big game you saw, made being in the wild a soul-enriching experience.

Sheila's son-in-law, Dallas, took us into the mountains above the farm. The Land Rover nosed through swollen rivers like an amphibian. We left the 4x4 and rock hopped into ravines lined with waterfalls and cave paintings. Looking at the human figures and animals, art from so long ago, I was filled with a similar sense of wonder to when we visited Lascaux II with Jim and Sheila years before. Lascaux is a major prehistoric cave painting site in our region of France. Lascaux II was a copy of the original, but we still sensed the magic.

Here the paintings were real, and the environment was wild, as it was when the people who made the paintings inhabited it. Nature untethered and precious. Following the ravine from the caves covered in paintings of hunters and antelope, we passed behind a waterfall. Shimmering curtains of water sprayed refreshing mist back at us. Through them, the hills of green, gold and ochre, moved like living things. I felt suspended in time and space like I was floating in the aboriginal 'everywhen'. As I write this, years later, I feel a great welling inside, at the spiritual joy of that place.

Back at their home farm, we walked down to the river to swim. The week before, parts of South Africa had experienced more rain in two days than they had in two years. We reached the river and stepped out into its cool depths. Sean's t-shirt bubbled up around him and we laughed. I lay back and the river carried me swiftly forward. Surprised, I pressed my feet back down and got a footing near the bend in the river. I waded back to the laughing crew.

'Wow, it's fast,' I said.

'*Ja*. We must respect it,' said Dallas. 'One of my dad's brothers drowned here when he was a toddler.'

The river took on a sombre note and I anchored my feet firmly to its bottom. The place was full of memories for Dallas and his family. As we walked back, Dallas told us that while farming in general was tough, the sheep business was doing well. They had created a system to raise sheep semi-intensively, and the relatively difficult access to their farm meant they didn't have as much theft as some farmers.

Sharing those precious days with Sheila, in the roundhouse she and Jim had built together, I regretted not having had the opportunity to know them better. As we left, gratitude for these moments with Sheila, her family, and their farm, in all its complexity and beauty overflowed inside me. It felt like I was leaving somewhere I was connected to, not somewhere I had just encountered.

Between potholes and road works we edged towards Haga Haga, a coastal village in the Eastern Cape. Each section of alternating circulation for the roadworks promised a wait of twenty minutes. Cars disgorged people to pee on the roadside. It felt a bit like home. France is known for informal pit stops.

The alternating circulation points were manned by two people, one on a walkie-talkie, and another to move the barrier. At our second stop, the barrier mover, bored senseless with his job, decided to play a game of chicken with the first car coming through from the other side. The oncoming car slowed but kept coming as the young man continued his game, pushing the barrier into the road and then pulling it away. At the last minute, he pushed it in front of the car and then pulled it aside with a bullfighter flourish. Inside our vehicle, there was a shared intake of breath at the closeness of it.

The SUV stopped a few metres past the barrier, directly opposite us, with four smartly dressed people inside. The driver leapt out and strode towards the matador, fists raised, yelling insults. I heard '*bulala*'; 'kill' several times. At the first, I locked the doors. At the second, I prayed we would not witness a murder. My Xhosa was limited, perhaps he was saying 'You could have killed us' rather than 'I will kill you', but I took his raised fists as a bad sign and gripped onto the car door like I did when I felt unsafe in a car going too fast.

'Relax Caro,' said Sean picking up my cue. 'Nothing serious is going to happen. Maybe a little fistfight, nothing more.'

The matador backed behind his large plastic barrier. The driver gave another volley of verbal fury. Everything went silent and the spectators in the queue of cars leaned in for the finale. Would he punch the guy? Pull out a knife or a gun? For a moment he appeared to be weighing the options, then he dropped his fists, got back into his car, and took off in a cloud of dust. We shared a collective outbreath of relief. Sophia and Ellie were wide-eyed, mobile phones down. At least we weren't getting bored at the roadwork stops.

We arrived at Haga Haga and after affectionate greetings and unpacking, Mum took me aside. Dad was forgetting things and couldn't do things that should have been easy for him to do.

'You guys are tired. Two months of constant travel, no settling down. I'd be losing it if I had done that,' I said, avoiding her worried eyes.

But I could see he had slowed. I hadn't seen my parents in years and the change was marked.

Dad and I walked to the grass edge of the sea. He used to hop across the rocks in front of us, agile as a mountain goat. On our holidays here, he was the most relaxed I ever saw him. He would spend days out there, alone with his pipe, fishing rod, and the wild sea. We turned and walked slowly together, following the path to the beach, and chatting about their trip. The rest of our clan disappeared into the distance. Halfway to the swimming area, a 20-minute walk away, Dad stopped.

'That's enough for me. I think I'll turn back,' he said.

'Okay, I'll see you back at the house,' I said.

I watched him take off slowly the way we had come, then turned and ran to catch up with my mum and my daughters. I found them searching for shells along the shoreline. We meandered enjoying the sound of the waves crashing on the rocks, the iodine scent of the sea, and the peace of this ancient activity. Shells were collected for jewellery and even used as money in ancient times. I remembered finding all manner of shells, including cowries and a mega diversity of other shell life along this stretch, as a child, forty years before. Now there was less diversity and fewer of them.

A factory farm for abalone that generated much-needed local employment had been built about a kilometre up from the beach house we were staying in. Wastewater reeking of chlorine rushed down the discharge outlet channel to the ocean when we passed by. The rocks near the exit were nude and some were black.

Researching the site as I wrote this book, I discovered that a major expansion was planned. According to a draft scoping report released in 2020. The Wild Coast Abalone Expansion, at Marshstrand, Eastern Cape, would increase the volume of seawater to be pumped from the ocean to the farm from 9 000 m3 per hour to 35 525 m3 per hour[11]. I wondered how much chlorine that meant would be washed back into the sea. Abalones were a premium luxury product. Their value was so high they were transported from the site to the airport under armed guard. From there they made their way to the Asian market.

At the beach we swam in the choppy Indian Ocean, wild wind whipped up waves where there were none. It was too windy to stay for long. Sea swimming was always a tonic, even in conditions that weren't ideal. On the way home, the sand stung our legs. We covered our faces with scarves to protect our eyes and noses. Back at the house, relieved to reach shelter, we found Dad sitting in an armchair, visibly shaken.

'I fell on the way back and couldn't get up,' he said. 'A woman in a pick-up truck saw me and came to help otherwise I would still be there. I'm fine now.'

'Too much whisky for breakfast Dad,' I said, trying to make light of it, but scared inside. This time I met the worry in Mum's eyes.

That evening, gathered with our friends whose family owned the beach house, Dad described how he had fallen into a bramble bush at home before they left. His legs gave way, and he went in head-first. The way he told it, it was funny, and we laughed, but inside I worried.

The seawater cured the remains of my eye infection, and I began to feel better than I had in months. Sean, Sophia, and Ellie left to catch a flight from East London. They had to get back to France for the start of the school term. I stayed on to accompany Mum and Dad back to Durban. We took the long road to Maclear, via Queenstown, thinking it would be faster, safer, and free of roadworks.

In the Thomas River district of Cathcart, undulating light green *veld* rolled into the distance, familiar territory, that brought reminiscences floating up from the recesses of Dad's mind.

'The hail stones were bigger than golf balls, some were the size of tennis balls. They had jagged edges and spikes. They killed sheep and cows, and smashed roofs and windows.'

His memories poured out as we drove. I was sure this one was well-embellished.

'Dad got his sheep into the barn but another fellow's sheep were pasturing nearby and there wasn't enough room for them. The hail killed them. There were dozens dead. The vultures arrived so fast you wondered how they knew. My Dad thrashed chains to keep them away from the carcasses so the farm workers could skin the sheep. At least the farmer would have those. Then, after everyone had taken what they could reasonably eat, it was over to the vultures. After feasting they were so heavy that when they took off they barely made it over the house. We could smell the stench of them from inside. I'll never forget it.'

'Incredible,' I said.

Now vultures were a rare sight. In South Africa, they were being wiped out by cheap poisons used to kill predators, like foxes and wild dogs. Vultures ate the carcasses, and they too died. In India, vultures were almost wiped out by diclofenac, a veterinary anti-inflammatory drug for pain relief in cattle, highly toxic to vultures. Ninety-five percent of their vulture population was gone by the time a ban on the product turned the tide. The disappearance of vultures led to a rise in other scavengers like rats and feral dogs. In India, the rise in feral dogs from the loss of the vulture population led to a rise in rabies cases for humans and thousands of deaths.

'Avondale's over that hill. We lost that farm when I was a few years old, then we moved to Topsea where the hail incident happened.'

Dad's parents lost Topsea too. His childhood was scattered with lost farms, each one imprinted in his memory. Perhaps it was part of why I had fought so hard not to lose our farm in France in our early days, the story told in my first book 'Grape Expectations'.

Beyond Stutterheim the road was wide and easy, perfect tar cutting through grassland that continued to the horizon. After Queenstown, we headed north, then east along the foothills of the mountains that circle the southern end of Lesotho.

With no warning, the perfect tar ended. Potholes that could break an axle, roadworks to fix them, and itinerant cattle, sheep, and goats, hindered our passage. This route had even more roadblocks than the shorter route, but it was too late to change now. The stress on Dad was visible and I took over driving.

After an eight-hour journey that should have been five, we picked our way through Maclear's streets, tumbling with street stalls, pedestrians, and buckled cars that clearly drove the road we had travelled regularly. On the long dirt road to our friends' house, the car slid, sashaying like a drunk. I geared down following Dad's guidance to weave the steering wheel counterintuitively to keep to the track.

Pip and Marie bounced out of the front door as I came to a stop.

'Come on, let's get your things in, then we'll put the car in the barn,' said Pip, giving us a welcome hug.

'I'll make tea then we'll go for a walk,' said Marie, as she hugged us hello.

Soon we were sitting in the lounge sipping strong African tea. Pip got up and went to look out over the veranda. A bank of dark clouds was growing on the horizon.

'I wonder if that storm is going to bring hail,' he said, then sat down again. He was relaxed. He and Marie had sold their surrounding farm a couple of years before. They had the use of the house for their lifetimes but no crops to worry about.

The trees began to thrash. The patter of rain on the roof became a drumbeat. The swimming pool surface started to dance, hailstones created flares, mini fountains that progressed from a slow waltz to a frenzied samba.

Leaves scattered down scythed by the relentless onslaught. The garden turned white as if it had snowed. Hailstones bounced up onto the veranda, bigger than golf balls and some almost the size of tennis balls. Perhaps Dad's story about the sheep wasn't embellished. We watched enthralled then realised closing windows and doors would be a good idea.

In my room, water seeped from the sunroom towards the carpet. I thought a window had been left open but seeing none, I looked up. The roof of the sunroom extension was smashed, tennis ball size holes open to the dark grey sky scattered the surface. Hail and rain poured in.

I yelled for help. Marie looked in then ran for towels. The hail was already centimetres deep and rising.

'Our roof is leaking too,' shouted Mum.

Marie ran for buckets for my parent's bedroom where the weight of the hail had undermined the roof, then she came back to me with more towels.

'With all the glass shards, don't come in without shoes,' I said.

Marie placed more towels, her feet bare.

'We need buckets and a spade to get the hail out before it melts and floods the house even more,' I said.

'Good idea,' said Marie as she continued to lay towels barefoot.

Marie returned with more buckets and two spades for the hail. Pip and I scooped, Mum mopped, Dad wrang towels, and Marie ran the relay to empty the containers. The hail stopped. We slowly got the upper hand on the flood.

Marie assured us everything would be okay. It looked like clear skies for the night. They would find a solution to the broken sunroom and the roof the next day. Now it was time to relax with a glass of the Feely wine I had brought and dinner.

The next day we stopped in for tea with other friends, as we made our way to Durban. The fields of maize around their place got us talking about agriculture. Nate explained that most of the grains sown in the district were genetically modified and coated with insecticides and fungicides. Farmers saved more than double if they used coated seeds. The birds used to eat half, or more, of the seed that was sown. If it was coated, they didn't. In my mind I wondered 'Or perhaps there were fewer birds because they died after eating the poisoned grain'.

Later, I looked up seed treatment. Research showed the coatings were dangerous to birds and mammals, and to the insects that pollinated the crop that grew from the coated seeds. Birds and mammals ate the spills, and if the sown seed was visible, they ate that too. The pesticide dust from the coating flew into the air during the mechanical seeding process, poisoning those in its cloud. Closed tractor cabins, with strong filtration systems, like the ones required for using the systemic chemicals in conventional winegrowing, were required.

The companies that sold treated seeds recognised the risk and put labels on the seed bags that said seed treatments were toxic to wildlife. They advised farmers to clean up spills or cover them with soil. But it wasn't only spills that harmed creatures. In a quarter of the fields studied, sown seeds were exposed on the surface and thus creatures ate them. Chickens, relatively large compared to the wild creatures concerned, would get drowsy, pass out and even experience muscular seizures when fed the seed.

Despite all these warning signals, for the year 2021, multiple research reports estimated the seed treatment market was already worth several billion US dollars. It was being filled by big agricultural chemical companies like Bayer and Syngenta. The technology had snuck into our food supply without people being aware of it. I had never heard of seed treatment before. I could see that it was tempting to farmers as it helped reduce their costs since they lost less seed. For the end consumer, the only way to avoid poisoned seed and their effects was to buy certified organic.

By the time we reached Steph and Dave in Durban, Dad was leaning slightly to one side when he walked. He couldn't find the light in their bedroom even though we had shown him where it was. Mum and I were worried, but there was nothing we could do so late in the day. Doctors' rooms had closed for the night. I was leaving early the following morning.

'As soon as you get to your friends after dropping me at the airport, you must book Dad to see a doctor,' I said.

'I know Toots,' said Mum.

At the airport, Dad's lean was almost 20 degrees, and he was shuffling instead of walking. He assured us he felt fine. It felt wrong to leave them with Dad in that strange state.

'Let me know what the doctor says,' I said as we hugged tight goodbye, tears pricking my eyes.

At Charles de Gaulle I checked messages. There was no word. It had been good to see our families and Dad's health problems made me even more appreciative that we had made the trip. The hail in Maclear and Dad's story showed me how dangerous hail could be. Like other climate extremes, hails storms and their intensity were increasing with climate change.

Back in Saussignac finishing the pruning kept my mind off the worry about my Dad. The frenzy of preparing for our annual bottling grabbed my attention. I was thrown back into life at Château Feely. Nature and the seasons dictated our workflow. They stopped for no one.

CHAPTER 8

Harassment

B ottling concentrated the mind and helped to forget everything else. It was like an ICU. Bleeping sensors monitored the bottle filling, the lifeblood of our enterprise, alarms sounded if there was a problem with any of the vital signs. Sincérité, Générosité, Sensualité, Liberté, Résonance, La Source, Vérité, Grace; the wines ran out of the vats and into bottles ready to tell their story of the vintage and the farm to those who would drink them.

Each wine had a personality that we had tried to capture in the name. Sincérité was Sauvignon Blanc grown on limestone. It was clean, clear, linear, and direct, like sincerity. Générosité was barrel-aged Semillon, a dry white wine that was rich and round, with depth, complexity, and a generous sensation in the mid-palate, like its name.

Everything was running smoothly, almost too smoothly. We were near the halfway mark when the forklift got stuck in the gravel. Sean ran for the tractor. Bruce, our brother-in-law, loped after him. They didn't want to stop our flow. We were under pressure to get it all done in one day. Minutes later Bruce leapt onto the forklift and Sean began pulling gently with the tractor.

I held my breath trying to stay focused on the line. One moment of inattention and I would lose pace and upset everyone. It was hard not to watch the forklift drama. Ludovic, the operator of the bottling machine, wasn't showing any signs of stopping for it.

Ian, a close friend, and I exchanged wide-eyed smiles and kept the bottles moving.

'The machine has three settings super macho, macho and wimp,' shouted Ian. 'It's on wimp for us.'

Our laughter resonated over the clinking bottles, the grinding of the compressor and the general hiss and hum of the unit. Ludovic gave us a look that made me feel like we'd been caught whispering in prep at boarding school. Ian and I cracked up. The finished boxes were piling up. Soon boxes would fall off the conveyer if Sean and Bruce didn't get back to stacking them. It was finely tuned timing with no leeway for stuck forklifts.

From the corner of my eye, I saw the forklift lurch out of the muddy gravel and felt a shot of relief. Sean parked the tractor and ran to lay cardboard on the sticky sections. Bruce moved the full pallet out of the way, and then they both raced back to stack the next pallet with the boxes jammed along the conveyer. Everything was back on track. We were exhausted and euphoric. Ian and I started laughing uncontrollably to release the stress. Sean rushed past to fetch the next pallet, sweating and red-faced. He gave us a Ludovic look that made us laugh even more.

Lunch was over in a flash, usually, we took time to be together and enjoy it, but we had to race on. By the end of the day, Sean had stacked and moved ten tonnes of wine. His body was sore all over. My wrists were strained from lifting each bottle of those ten tonnes and packing it into a box. While it had been fun, knowing how sore we were, I knew we couldn't expect our friend Ian, who was ten years older than us, to help in the future. We had been lucky to have Bruce with us. What would we do for the next year's bottling? Each year it became more difficult to find short-term labour. I put the thought aside, a challenge to face at another time.

Spring tour guests ramped up, bringing a cosmopolitan touch to our rural lives. A few days after bottling, a vibrant group joined me. They had finished the vineyard and tasting part of the tour and were well into the drinking and lunch part.

'What is it with the no seat on the toilets in Paris?' asked Huan from Sydney. 'What are we supposed to do? Am I supposed to sit on the ring without a seat?'

'Exactly! This is France, right? In Singapore we buy toilets 'made in France' that are beautiful, hand painted ceramic art works. We get here and... not even a seat,' responded James.

Guffaws echoed around the room. James and his partner were architects from Singapore. They designed top-end hotels and resorts. With around 80 million visitors a year to Paris at the time, and with tourism a major driver of the French economy, it seemed the least we could do was put a seat on it. I was laughing but also a little embarrassed.

'What about the toilets that are holes in the ground?' asked another guest. 'There are still lots of those in France.'

'The idea is to squat so that you don't stay for long,' said Body, a Nigerian living in Bordeaux. 'We don't want you to settle in with your newspaper. We want you to do your business and get out.'

There was another burst of laughter. It wasn't only public toilets that could up their game. Our place needed attention too. The Wine Cottage self-catering apartment was undergoing renovation. Our kitchen was next in line. For ten years, we had squeezed our family into two small bedrooms, a shower room, a lounge, and a tiny galley kitchen. The year before, Sean placed two armoires down the middle of Sophia and Ellie's room, as dividers for privacy. It was a gesture, but they needed their own rooms. My soothsayer, Lijda, had foretold it on their visit the year before, and, as usual, she was right. To that end, The Wine Cottage renovation was decreasing the Cottage floor space so we could reclaim a room for Sophia.

That afternoon I decided to break the habit of not walking with my daughters and went up to meet them at the bus. Dora needed a walk and so did I. As we walked back, I noticed Sophia was agitated.

'What's wrong my lovely?' I asked hugging her as we kept walking.

'A boy in my class sneered at me for being English,' she replied.

'Oh Fia, I'm sorry,' I said. 'He's probably jealous that you do well in English.'

'I do well in French too, so that's no excuse,' she said as her face crumpled in tears.

'Oh, my lovely,' I said. 'I'm so sorry.'

Sophia's story was a shock. We hadn't come across this sort of harassment before. We generally felt welcome in our community. But we had heard other stories. One of our winegrower friends had been harassed. He and his family were taking a quiet walk on their farm one Sunday when they came across two hunters.

'*Messieurs*, this is a no-hunting area because of guests and young children,' said Dom politely. 'You can see the signs.'

The hunter came up close breathing alcoholic fumes onto Dom's face.

'I'll hunt where I like!' he shouted. He started walking away then turned, shot into the air above their heads and shouted, 'We'll get rid of you English, get you out of France, once and for all.'

Dom, his wife, and their young daughters were deeply shaken. When I heard the story, I was amazed that they had not pressed charges and horrified that any French people felt this way. Dom wanted to avoid taking the conflict to a higher level since they knew the people were locals. Instead, they met the mayor about the incident, but as far as they knew no action was taken.

We had been fortunate, the community of Saussignac was welcoming to newcomers and non-French nationals. Sean had had a few words with hunters on our land, but never anything like this. Most of the hunters were respectful of us, as we were of them. He had been threatened once, not about being foreign or about hunting, but about farming.

We had to do organic anti-fungal contact sprays before the rain. If we didn't, fungal diseases like downy mildew could destroy our fruit and even kill our vines. We couldn't wait until our vines were sick, then spray them with a systemic fungicide as chemical farmers did. We needed to do preventative sprays before the rain, and before it got hot, otherwise the leaves and fruit would be burnt. There was no point in spraying if we were not certain that rain was coming, as it would serve no purpose.

One Saturday evening Sean's weather app predicted a storm for the following late afternoon. It was a while since our previous spray so he couldn't risk leaving it. He had to spray. Sunday morning, he started on the vines furthest from the houses around 5 a.m. He finished and was inside for breakfast by 10 a.m. having cleaned the spray machine and tidied up. We heard an aggressive bang on the door. Sean opened it.

'Why are you spraying outside my house so early on a Sunday?' demanded a neighbour.

'I had to spray. A storm is forecast for tonight. If I hadn't, we would lose our crop,' said Sean.

'No way. Look at the weather. It's beautiful. It isn't going to rain,' said the man aggressively. 'Anyway, you could do it at a civilised hour, especially on a Sunday.'

Sean calmly explained that if we sprayed after ten in the morning, we would burn the leaves and the fruit, especially with the hot weather forecast to precede the storm. To get around all the vines before ten he had to start early. He was interrupted before he could add how important it was to spray early to avoid upsetting pollinators like bees.

'It won't rain,' said the angry neighbour. 'Anyway, I don't care what the reason is, you can't wake us up so early on a Sunday. Next time I'll bring my gun.'

He stomped off before Sean could say another word. Sean closed the door and slumped into his chair. He had to get up early on a Sunday morning to work. Now he had to deal with a gun threat for protecting his livelihood. I was

upset and Sean was bemused. It didn't make sense. Our
neighbour should have been thanking us for being organic
and saving him and his family from pesticides that were
carcinogens, nervous system disruptors, and endocrine
disruptors. Instead, he was offering threats.

That evening a storm accompanied by a massive
downpour blew in. I felt like going and banging on his
door. Instead, I poured my harassed heart out to Isabelle,
a friend who was a teacher and a winegrower, on our trip
to our weekly yoga class.

'Yes, we can find anti-English sentiment at school. And
beyond school too. Especially around Sophia's age, kids
can be mean. You must go and see Sophia's teacher.'

At yoga, I felt my worry ease away as we followed an hour
of slow, relaxing poses. Yoga was medicine for stress.

I took Isabelle's advice and booked a meeting with
Sophia's teacher. But by the time we got to chat on the
phone, Sophia had sorted the problem out herself and
didn't want me to say anything. I felt upbeat. Sophia had
solved her problem. The spring growth looked good; lime
green shoots were dancing like fairies on the canes.

The next day mown areas of the vineyard were white
with frost and crunchy underfoot. The young Cabernet
Sauvignon on the exposed plateau was badly hit. After a
walk around the farm, we estimated we had lost about half
our harvest. We would find that it was worse. It was bad
news given our fixed costs would not change, we had to
keep farming even if the vines didn't fruit.

We hoped that new shoots would grow, but they
would be so far behind it was unlikely they would reach
useful ripeness. Fortunately, we bottled our wine and
commercialised it ourselves so we could smooth the loss

over a few years. We'd feel the initial hit with the Sauvignon Blanc and the rosé. They were wines we bottled six months after harvest. The loss on the barrel-aged white and the reds would be felt over several years, as they would be bottled eighteen to thirty-six months later. For wine farmers who sold in bulk, the loss would be realised in one go.

Longer term it was a bad sign. When we arrived in 2005, wine growers in the region expected extreme weather, like frost or hail, to hit them once every ten years, now it was three times every ten years and rising. We were right on that new average, having been hit by one hail event and two frosts in the ten preceding years. Global warming wasn't only warming, it was also extreme weather events. As farmers, we were feeling it up close. I thought of Sandrine, our colleague who had lost an entire harvest, and got divorced as part of the fallout, and felt a shot of adrenalin. I sent a prayer to the universe asking for protection for my farm and my family.

Sean stopped ploughing the vineyard. The permanent plant cover stopped erosion and offered natural insulation and biodiversity. To keep plant growth in check, he mowed infrequently between the rows, used a mechanical hoe or a strimmer under the vine rows, and hand-weeded the young vines. We have thick mats of clover in the spring and more biodiversity from not ploughing. Not ploughing saves tractor hours, fossil fuel, and the earth's skin.

My third book in the vineyard series, 'Vineyard Confessions', hit the shelves. On my way to a radio studio in Perigueux, the capital of our department of the Dordogne, for an interview with BBC Radio 4, an oncoming car side-swiped me. With no safe verge, I had nowhere to go. I felt the shock of the impact and pulled over my heart racing. The perpetrator disappeared into the distance.

My wing mirror was smashed off, but the rest of the car was fine. It was so violent I thought there would be more damage. I sat for a few minutes, my hands shaking, debating whether to go home given the shock I was feeling. After a couple of minutes, I decided to continue. I arrived at the interview with seconds to spare.

Back at home after the radio show, it felt like the incident was sign of wider collapse. Things were falling apart: my relationship with Sean; the planet and climate extremes manifested in the late frost; my dad's undiagnosed illness. I told myself to focus on the positive and to remember everything I had to be grateful for. A roof over my head, food on the table, incredible daughters and a kind husband. Perspective changed everything. As I packed my bag, I had sparks of happiness in my belly merely by changing my thoughts. Nothing had physically changed from ten minutes before when I felt like the world was caving in. All I had done was change my frame.

I was on my way to Ireland to speak at Ballymaloe's International Litfest. The festival was about being literate in food and its provenance, and a celebration of food and wine related books. It was the perfect destination to lift my spirits, to promote 'Vineyard Confessions' and to participate in the Ballymaloe Litfest team's mission to 'inspire positive change in the world of food'.

CHAPTER 9

Inspiration Fest

B allymaloe is a household name in Ireland, synonymous with great food. Ballymaloe House started as a restaurant. The founding chef, Myrtle Allen, created simple real food from raw ingredients that came direct from the family farm, local growers, and fishing boats in the nearby seaside village of Ballycotton, a 'farm to table' experience before it was fashionable. The restaurant developed into a hotel to get a liquor licence. At the time, liquor licences for restaurants and bars were expensive but they were easily available to hotels. The family renovated ten rooms of the country house into hotel bedrooms, the number required to be considered a hotel.

It is now a sprawling success story including farm, hotel, restaurant, event spaces, shops, cookery school, branded produce, and famous chefs. Perhaps the best known Ballymaloe personality is Darina Allen, who arrived to work as a trainee chef in the Ballymaloe House kitchen and fell in love with Tim, one of the Allen's sons. A few miles down the road from Ballymaloe House, she created the internationally renowned Ballymaloe Cookery School, with her brother Rory O'Connell. Darina became a legendary chef and cookbook author. More recently, her daughter-in-law, Rachel Allen, brought the next generation into the limelight, as a television chef, and cookbook author. The greater Allen family had spawned many food businesses and personalities over the years.

Ballymaloe serves Feely wines in the restaurant and at the cookery school. We had worked with them to offer events, including wine and food dinners, over the years. This was my first time participating in their International Litfest. The programme included multiple events running concurrently in different rooms at Ballymaloe Cookery School and Ballymaloe House, and a fringe festival. Several thousand people were expected to attend. I looked out of the sash window in my room as a flutter of excitement swished through me. Below, a walled vegetable garden connected to fields and hedgerows that ran to the horizon. The place exuded peace and timelessness. Right then, timelessness was not for me, I was late for the welcome dinner. I applied lipstick, then skittered down the stairs, and across the courtyard.

Ballymaloe House Restaurant was a series of connected rooms in an old Georgian house, part of the original family home. High ceilings, white linen, and wood chairs with velvet seats, float over plush carpets. The rooms were packed with food personalities, chefs, farmers, writers, film producers, journalists, bloggers, and our Ballymaloe hosts. They buzzed with excited exchanges on food, wine, and farming. I met the producer of a Netflix series, two chefs who had recently launched their first cookbook, the McKennas of McKenna hospitality guidebooks, and an investigative journalist, Joanna Blythman. The evening was full of hope and ideas for healthy food and a healthy environment.

In the morning, at the first panel discussion with a local organic farmer, a community gardener from London, and a farm coordinator from New York State, I was brought back to reality.

'Only two percent of Ireland's farmland is organic,' said the local farmer.

We had experienced going from 1 percent organic to 10 percent in French vineyards. At 1 or 2 percent, organic farming was a tiny niche. At 10 percent it was taken seriously by the Department of Agriculture and other support services, and there was a snowball effect.

Listening to them, my mind wandered to how we could harness community and culture to raise awareness of environmental questions like climate change and farming. Both were key to our future but for most people they were 'boring' subjects. How could they be made mainstream? Ballymaloe was doing it with their festival that went beyond the speaking events to offering workshops on 'Grow it Yourself', cookery demonstrations, and information on composting and waste, but also dancing, and fun. Art in all its forms was needed to address the environmental challenges of the 21st century.

The core of Ballymaloe was a working farm. During the Litfest they transformed their massive tractor barn into a 'big shed' of artisanal producers and a night-time party hub. I bought lunch from 'Fused', Japanese food with an Irish twist, and made with local ingredients; then found a spot at a long trestle table packed with people. The person opposite me was solo so we got chatting. She had authored a guide to starting your own Small Food Business in Ireland. She was fired up about the quality of Irish food and the potential for small food businesses. We swapped cards. I wanted to linger but I was booked to hear Isabelle Legeron, nicknamed 'the Crazy French Woman', in the Drinks Theatre.

Isabelle was the first French woman to become a Master of Wine (MW). She made her name as a proponent of, and specialist in, natural wine. She founded RAW wine fairs, starting with a London edition. We attended in the early days when we were still attending wine fairs. Since then, we had built our direct sales to the point where we no longer needed to do fairs or '*salons*' as they were called in French. Isabelle's network of RAW fairs had grown into multiple international events including London, Berlin, New York, and Los Angeles. She had recently launched her book 'Natural Wine: An Introduction to Organic and Biodynamic Wines Made Naturally'.

Isabelle took the stage. She had cropped dark hair, a gallic nose, and a straight-talking manner.

'Don't expect the aromas and flavours you were taught to expect for varietal A or B. Taste the wine for itself, not for what you think it should be,' she said in a lovely French British accent.

'But I think this wine's faulty,' said a member of the audience. 'It's fizzy.'

'Perhaps it is fizzy. That doesn't mean it's faulty. Taste with your gut not your head. Taste as if it's a new drink, something that doesn't have to fit into a prefixed box for its grape or its appellation,' said Isabelle.

We tasted a series of intriguing wines. My favourite was a red and white blend called 'Jumping Juice' by an Australian winemaker. It was a blend of red made like a white wine, and white made like a red wine. It had vivacity and energy. I loved it. Isabelle's session confirmed how exciting natural wine could be. They took us to new places in our experience of wine.

That evening I introduced Feely wines paired with a pop-up dinner by Edinburgh Food Studio, a food research hub and restaurant created by a Ballymaloe alumnus, Ben Reade, and his partner. Their restaurant in Scotland offered a four-course tasting experience described as a 'real-time creative response to seasonal ingredients.' Our Irish importer, Mary Pawle Wines, had made sure the right Feely wines were lined up. Mary and her husband, Ivan, were deeply embedded in the food and wine scene in Ireland. They had been importing organic wines into Ireland for more than twenty years.

The seven-course feast was a voyage for the taste buds that included two highlights, turbot & lobster with Feely Sincérité pure Sauvignon Blanc, and hogget leg with Feely Grace red wine. The hogget, a one- to two-year-old lamb, too old to officially be called lamb and too young to be called mutton, was extraordinary with Grace. Comments from guests included 'a delight for the tastebuds', 'the best pop-up dinner ever', and 'fabulous wines.' I floated out on a cloud of happiness.

Sunday morning, John McKenna of McKenna Guides, and Joanna Blythman, industrial food investigator and author of 'Swallow This', and many other books, bounced off each other like professional talk show hosts. I had met them at the welcome dinner, and they felt like old friends. Litfest brought people close. Perhaps it was the shared dinners and breakfasts, the shared passion for healthy food, the general ambiance; most likely all of these.

Joanna's message was 'We need clean labels', that tell the full story.

'The 'Big Lie', that processed food is as good as real food, must be debunked,' she said in a fabulous Scottish accent. 'We're bombarded with beautiful labels promising wonderful food from 'Sunnydale Farm', only to realise that it is an intensive factory farm where there is no sun, no dale, and no 'farm'.'

Laughter rippled around the room.

She told stories of going undercover to industrial food fairs where only professionals were allowed. She saw products that could extend shelf life at the cost of nutrition, high-tech solutions for more convenience, more speed, more plastic, and the horror of intensive animal farming. I remembered my conversation with the Ph.D student on the train from Strasbourg and wished I had his contact details to send him a copy of Joanna's book.

Litfest hosted speakers with strong messages like Joanna and invited visitors that included supermarket executives and buyers, and large-scale food producers. The Ballymaloe Litfest was brave. It got people talking from all sides of the food and drink business by creating a place to share ideas.

My main speaking event was in conversation with Tomas Clancy, wine correspondent of the Sunday Business Post. At our first event together, years before, he started by saying he wondered why people like me moved from Ireland, as he showed a photo of someone leaning into gale force wind and pelting rain, to France, as he switched to a photo of Feely vineyard bathed in sunlight. I still laugh thinking about it.

At Litfest, he asked us to imagine where Château Feely would be in 300 years, after citing the 300-year history of Château Lynch Bages, which he pitted alongside Feely Grace in the comparative tasting. Lynch Bages was more than ten times the price. Grace stood up well and I felt proud. We tasted Feely *Sincérité* Sauvignon Blanc and a Feely *Premier Or* Saussignac sweet wine. With each wine, I told part of our story and touched on an element of our ethos and what organic and biodynamic farming meant.

'How long will these low sulphites and no sulphites wines last?' asked a member of the audience.

'Our first no sulphite red is nearly ten years old and is still in great shape,' I said. 'Age worthiness isn't created by the level of sulphites, it's created by the life force, the natural antioxidants of the wine, and elements like tannin and acidity.'

The audience was warming up and eager to ask more questions, but we had to give up the space for the next event. Reluctantly we left the stage, and Tomas and I said affectionate farewells. I didn't know it would be our last. When Tomas departed the earthly stage, a couple of years later, I wept. He was charming, witty, and kind.

After waving goodbye to Tomas, I realised I could make a talk by Christian Puglisi, whose restaurant, Relae, in Copenhagen, was the first Michelin-starred restaurant in the world to be certified organic. I legged it to the seminar room noted in the catalogue. The event had already started. I settled quietly into a seat at the back.

'High-quality ingredients must be produced naturally, that is, organically,' said Christian. 'The only way to be sure of the provenance, unless you are the farmer, or know the farmer and farm well enough that you do your own 'certification' visits, is with certified organic.'

I had found someone who understood as deeply as I did, why organic farming and organic certification were important.

'For the level of organic certification that our restaurant has, a minimum of 90 percent of the products and ingredients must be certified organic. That's a challenge,' said Christian.

I knew how difficult a challenge from pairing lunches at Château Feely. We aimed to cater 100 percent organic, to match our organic ethos, but there were times when we couldn't source organic versions of products we wanted.

'We had to adapt the menu to the availability of organic produce,' he continued. 'But I got frustrated with the lack of certified organic ingredients and not being able to serve what we wanted to serve. That led to us creating our own. We started a small organic farm for vegetables. Then, to be able to offer fresh organic cheese in the style of mozzarella, we added a small dairy herd and made the cheese ourselves. Having a farm, brought me and my team closer to the ingredients, and their provenance. We experiment with completely different vegetables and heritage varieties that are not available from our suppliers.'

Christian had worked for some of the most famous restaurants in the world, Taillevent in Paris, El Bulli in Spain, and Noma in Denmark, before creating Relae.

'In the beginning, Relae was not organic,' said Christian. 'The arrival of my son made me rethink my ingredients. I wanted to feed my son organic food. But we were a small family business. My partner and my son ate with the staff at the restaurant. How could I feed my son organic but not my staff? Or my clients? I realised that if I felt it was important for my son to eat organic, I needed to put organic on the table for everyone. That pushed me to find new suppliers, to innovate with the menu, and eventually, to create the farm.'

In his book 'Relæ: A Book of Ideas', he writes that gaining organic certification for the restaurant was one of his proudest moments. It was tough to meet the stringent requirements and associated red tape, but it made him ask his non-organic suppliers hard questions, that helped him see why they weren't certified organic, and hence why he was no longer happy to support them.

His sentiments resonated deeply with me.

Through Relæ's organic certification journey, and partly because of this journey, he expanded to create additional food and drink activities. He created a natural wine business to import the organic and natural wines that they wanted to serve in the restaurant. After going the extra mile to create natural organic bread from organic, locally grown grain, pure water, salt, and their own leaven, he created an organic bakery.

Christian's story was inspirational. Feely farm was proud to celebrate ten years certified organic at the time. We knew the commitment required, and how important it was for our customers, ourselves and our wider environment and community. It was good to hear from someone that was on the same wavelength as us. Being the *avant-garde* was never comfortable, but it was full of possibilities.

At the final dinner, I joined Isabelle Legeron, her partner, and Samuel Chantoiseau, the sommelier at Ballymaloe. We chatted about natural wine, tasted food and wine that set off explosions of happiness, then danced into the night in the Big Shed. The Litfest was an opportunity to dream big dreams about how to create a better world of food and drink. I had set the objective of doing more outreach, and more speaking on organic, and Ballymaloe had offered the opportunity to do that. It made me eager for more.

There was work to do. My friend Aideen and I visited a café and wine shop for tea and cake in Dublin the next day. I asked the manager if she could show me their organic range. After some reflection, she pointed out ten wines in their range of hundreds. On turning them around and checking the back labels, we discovered that only two were certified organic.

'Why is there no organic sign on the bottle?' asked Aideen innocently.

'Well, they aren't certified organic, but they are *lutte raisonnée*, which means they spray if they need to. Or some are HVE, which is as good as organic,' said the wine shop manager.

Before we had a chance to respond, she was called away for another client.

'That makes me so angry,' said Aideen. 'I see how hard you work and the sacrifices you make to be organic, then someone like this says wines are organic when they aren't. It's infuriating. You take the lower yields, longer working hours, and pay for your certification, then salespeople confuse customers like this.'

Aideen was right. But how many people realised this truth?

Supermarket shoppers often accept meaningless words like 'natural', 'sustainable', or 'green' , and don't investigate the details. As Joanna Blythman outlined in her book, we had to look beyond the nice words and colours on the label. You had to scratch the surface like Christian Puglisi did when he asked the hard questions of his suppliers.

'*Lutte raisonnée*' was not organic. It was 'I think before I spray systemic chemicals'. It required the farmer to assess the situation and decide how much to use, rather than spraying the dose of pesticide recommended by the manufacturer. I would expect all farmers to think before spraying toxic systemic chemicals on their farms.

HVE, or '*Haute Valeur Environmentale*', was a new certification created by the French government, and while it was a little better than '*lutte raisonée*', it still allowed the use of systemic chemicals. HVE offered a wide environmental picture and included four main elements of evaluation: biodiversity, *phytosanitaires* (treatments for plants, which include systemic pesticides like systemic fungicides and insecticides), fertilisation and water management. Under both '*lutte raisonée*' and 'HVE', farmers could still use systemic pesticides that were carcinogens (cancer-causing), nervous system disruptors, and endocrine disruptors (reproductive and hormone disruptors).

With pesticide scares and more consumer awareness about pesticides, some conventional winegrowers were tempted to say, 'organic but not certified'. Some said they didn't want certification because of the paperwork involved. But organic does not increase paperwork since the farm accounts and vineyard and winery inputs must be tracked by all farmers. The difference is, despite the danger of the chemicals used under conventional farming since there is no certification, this information, while required, is rarely checked.

Aideen and I settled into a cosy corner to share a slice of lemon cake and a pot of tea. It was a luxury to chat with a girlfriend. Aideen had recently taken early retirement from her job as a management coach in a large technology multinational. She had started a Master's in Spirituality. It was the beginning of a new phase of life.

The night before I left, we met with our dinner club group of friends at a restaurant known for its artisanal and organic message. Seeing no wines marked organic on the list I asked what was organic. The waiter recommended a Sauvignon Blanc from France, which I ordered. He brought the wine, opened it, and poured. When he left, I looked at the label and found no organic sign.

When the waiter next passed, I asked where the organic logo was. He looked carefully at the bottle and was shocked. The previous sommelier, who left two years before, had told him it was organic, so he had carried on sharing the error. He was super-apologetic and had a great sense of humour, so we laughed about it. But the truth was they had no organic wine on their list, despite the image they projected, and he was committing fraud. With organic rules part of EU law, if you buy a certified organic product, you are guaranteed no systemic pesticides were used in its production. Seeing the logo on the product is important. My daughters joke that I am a wine waiter's worst nightmare.

Saying goodbye to our friends, and to Dublin, was poignant. It still felt like home. I felt tied to it, attached to the constantly changing sea and sky, the familiar streets, and the people.

Back in Saussignac, my first task was a multiday tour for an English couple. At a St Emilion estate requested by my guests, the guide said they were 'organic but not certified'. I bit my tongue. The first Medoc *premier grand cru classé* to be certified organic, Château Latour, had recently announced its certification. High-profile wineries raised organic awareness and provided even more temptation to say, 'We're organic but not certified', like this guide had done.

I drove us to *Atelier Candale* pointing out famous vineyards en route. The restaurant was perched amongst the vineyards and surrounded by manicured herb gardens. Purple echinacea hugged up to blue-green fennel, pale sage, and fragrant thyme. The surroundings and creamy courgette soup topped with purple potato chips and parmesan crisp smoothed my ruffled edges. By the time I sank my spoon into a satiny chocolate mousse, the guide's lie was almost forgotten. I felt velvet soft, indulged by gorgeous food and vineyard views.

The only way to fix misleading information was to educate people about farming and food provenance like Litfest aimed to do. It was a melting pot of people that helped grow food and soul; and generated ideas and projects bigger than itself. Every moment of the multi-layered experience lived up to my initial feeling of excitement: speakers, food and drink, fringe events, gardening ideas and offline exchanges. It inspired me to reach out to a local Michelin star chef Vincent Lucas, of

'*Etincelles*', to create a new wine tourism offering of four '*bouchées*' or bites, created specially by Vincent to pair with four Feely wines.

My sister Foo booked the Wine Cottage at Château Feely for the summer offering a treasury of shared moments and an anchor income that helped offset the frost. We had time to chat, to share memories, dreams, and laughter. With Foo staying, I slept better and felt more in control. I didn't get sick. It was lucky since our GP had retired with no replacement. I tried to find a new doctor, but no one was taking on new patients. We lived in a medical desert. Until recently I hardly needed a doctor but given the previous year's bouts with gastro and infections, I wanted to find one.

Foo left. As autumn handed the baton to winter, another round of gastro hit me. While laying up in bed, I researched relationship wisdom and bought a book about how caring more for your kids than for your partner was a sure way to kill your marriage, and what to do to have a balanced family life. I read it and found many messages that made sense. I asked Sean to read it, but he didn't see the point. Tears formed as I put the book back into the drawer. We were losing our way. We both needed to pull our weight to get our relationship back and Sean's rejection felt like he wasn't. I needed to get away for some perspective.

CHAPTER 10

The magic of biodiversity

A biodynamic workshop offered a much-needed day away, at a cosy farmhouse, near St Emilion. Introductions passed around the room and I scribbled names in my notebook. Attendees included people with deep experience and newbies. Among the new were a Chinese woman who managed a large Bordeaux estate for investors, and a French man from a Margaux grand cru classé. In the ten years since we started practising biodynamics, it had become chic.

After administrative updates, the meeting turned to problems in viticulture, like the virus, *flavescence dorée,* spread by the American grapevine leafhopper. It was brought to Europe by human movement, perhaps during World War II; a story that would have been at home in Elizabeth Kolbert's book, 'The Sixth Extinction'. Two participants had grubbed up several hectares of vines because of it.

'*Flavescence dorée* brings death to places that are not good for vines, places that are wet and compact,' said Paul, one of the original biodynamic growers of the region.

'I think cloned vines are also more susceptible,' said Jean-Michel, technical director at a famous biodynamic estate. 'We create a perennial monoculture, then, we create a perennial mono-clonal-culture and what do we expect?'

Paul was like Father Christmas, calm and soft-spoken. Jean Michel was straight as a cane. They had decades of experience in biodynamic farming.

'State aid supports cloned vines only, not massal selection,' said a participant.

'Where can we get alternatives?' asked another participant. 'All the nurseries I know only offer clones.'

'I'll send the details of the one I know,' replied the host.

Sean and I didn't want to follow the cloned vineyard bandwagon for our new plantations. Everywhere we looked, new vineyards had sprouted packed with identical cloned vines. To receive plantation aid money, the French government required winegrowers to use specific clones, within specified varietals allowed in their *appellation*, or protected designation of origin. *Appellation*, or in EU English, protected designation of origin (PDO), rules and associated bureaucracy are a constraint to wine farmers adapting to climate change. The appellation specifies strict rules that range from what grapes you can grow, to how you prune the vines, and even how you make the wine. While PDOs protect tradition, they also act as a brake on innovation and adaptation. There is little freedom of choice or biodiversity.

Cloned vines are cuttings from a specially selected vine that are identically propagated, often over decades, some since the 1970s. Massal selection, meanwhile, takes diverse cuttings from old vines, that date from the era before clonal selection, and reproduces them. Every year there are fewer of these ancient *grandes dames* – great ladies – vines are feminine in French. At Feely farm, we are fortunate to have blocks of pre-clone Semillon and Merlot. Massal selection means better genetic diversity, and, some would argue, a better reflection of unique '*terroir*', or taste of place. Cloned vines mean standardisation in the vines and in the taste of the finished wine. At Feely vineyard, we aimed to create unique wines. We were not in the game of standardisation so massal selection made sense for us.

'There are more disease problems with fire varietals like Cabernet Sauvignon, planted in soil that has strong water and earth forces, like clay and limestone,' said Paul. 'In these conditions, bringing some fire, with silica, could help.'

In biodynamics, we sometimes characterise plants in terms of the four elements, earth, water, air, and fire. For example, Cabernet Sauvignon is a fire varietal that likes to be in a relatively hot environment with good drainage. It typically creates high levels of sugar, and thus high alcohol wines.

Picture clay and limestone compared to gravel. The clay and limestone will be cooler, there is more water retention than in gravel, which will drain water away. Gravels become hotter, and keep the heat stocked for longer, than limestone. Cabernet Sauvignon prefers the drained gravel environment. This is one of the key reasons why we find more Cabernet Sauvignon on the left bank of Bordeaux, where a layer of gravel and sand overlays the limestone. A glacial melt brought gravel from the Pyrenees mountains, millions of years ago. On the right bank, where there is more clay and limestone, more Merlot is planted, because Merlot likes the slightly cooler environment.

Grape varietals are changing with global warming. With changing climatic conditions, the crop, and the variety, that does well in a given place, will change. In 2021, Bordeaux approved planting of new varieties, Touriga Nacional, Castets, Marselan, and Arinarnoa, for reds, and Alvarinho and Lilorila, for whites. In 2021, Bergerac added Fer servadou and Mérille as accessory red grapes. At the time of writing, Marselan, Arinarnoa, and Saperavi – a Georgian variety resistant to disease and frost – red grapes were being trialled in Bergerac, but were not approved yet.

We knew what silica fire could do, from my experience with burning our Cabernet Sauvignon vines. The two key preparations in biodynamics are the 500 horn-manure preparation and the 501 horn-silica preparation. Despite their tiny doses – a small handful per hectare for the 500, and less than a teaspoon per hectare for the 501 – they are powerful preparations to be used with care, not products to apply by rote.

Sean was wary of the horn-silica preparation given the rude summers of Southwest France. During a cold, rainy spring, I convinced him to do a second spray of 501, taking the cue from a biodynamic colleague in Jura who saved his crop in a very wet year with eleven passes of 501. A week after the silica spray we catapulted from cold temperatures to a peak of forty degrees Celsius. A hot wind from the Sahara amplified the effect, making it feel like someone had a hairdryer blasting at us. The dose of powdered horn silica is minute, a couple of grams per hectare. We mix the silica in water in a special way, called dynamisation, then we spray that water in a fine mist, into the air above the vines. The dose is so tiny; it is hard to believe it has an effect, but it does.

The 501 magnifies heat and light forces. The day after the Sahara wind, I walked past our young Cabernet Sauvignon. They looked sick. Their leaves were partially brown, some were completely crisped up, dead. On that first terrifying glimpse, I did not connect the heatwave to the situation. I kept walking and talking, not wanting the group that was with me, to see how upset I was. We were following the same route I had taken to the *Festival des Ploucs* with my daughters. Beneath Saussignac castle, I drew away from the group to call Sean. I had not seen damage like this before. Fear jabbed in my belly. Sean went to look. He called back as we descended into the forest. The vines were burnt. It was the heatwave. It was bad, but he thought they would recover.

Perhaps they would have been burnt without the silica spray, perhaps the heatwave would have left a trail of destruction regardless. Still, I felt culpable for their pain. Sean left a silence across the ether stronger than an 'I told you so' look. When I returned that afternoon, I looked more closely. The leaves were burnt, like someone had passed with a flame thrower. Our other vines were more fortunate, not as exposed, facing east or north, rather than on the plateau; or they were older and their roots reached deeper into the cooling limestone. The older vines were also wiser. They had more experience of how to cope with extreme events than the fiery Cabernet, our youngest vineyard. Fortunately, the vines survived, I, and they, wiser for the experience.

'But be careful of the power of silica,' said a participant as if he had read my mind. 'We also need to plan more biodiversity. This helps to manage insect populations.'

'Animals are important too,' said the Demeter representative.

'Yes. We bought fifty cows, so we have enough manure to make our own biodynamic compost and horn preparation. We bought heritage breeds from the Conservatory of Aquitaine,' said the participant from the Margaux estate.

For the grand châteaux, budget was not an issue, unlike those of us of more modest means. Some growers had sheep grazing in the vineyard, sometimes their own, sometimes hired. Hired sheep were often trucked in, somewhat negating the ecological objective.

'Be careful of the breed of sheep,' said Olivier. 'You must choose one that will not eat the bark and kill the vines. There's a Scottish breed that many growers have recommended.'

'Our friends in Alsace have Rackas,' I added.

'We have geese,' said another grower. 'Indian runner ducks and chickens can also work.'

A debate about how to protect your fields from neighbouring conventional growers followed.

'You need two-metre-thick hedgerows to protect your vines from systemic pesticides,' said Thomas, a farmer from *Les Landes*, south of Bordeaux.

Hedgerows offer habitat and food for insects, birds, and other creatures. Linked hedgerows create invaluable biodiversity corridors across a landscape.

That evening, filled with a sense of community from the meeting, I shared what I had learned, with Sean. We spoke deep into the night about things we could do, about what to plant on the section of vineyard we had grubbed up at the foot of the hill closest to the district road. We both loved the Chardonnay wines we had tasted from our region. We felt ready to shake up the grape varietals we had. We talked about developing more hedgerows and getting animals. Despite what our friends in Alsace had said, it was difficult to see how we could include animals, beyond the chickens we already had, in our daily work.

The vine nursery mentioned at the meeting offered massal selection Chardonnay. The following day I got in touch. Although the cost was almost double a standard clone, and we wouldn't get aid for vines, we knew massal selection was the best way forward for quality, disease resistance and for genetic diversity. We hoped one day to make a massal selection of our old vine Semillon and Merlot but that would be another whole level of investment. As a client said earlier that year, 'It's like me with my woodwork. You buy the best wood and cry once.'

We had to commit to the order nearly two years before we would receive the vines. It was a long game. Trying to imagine what our microclimate would be like when they came into production seven years later was difficult, projecting more than forty years out, to when they would give old vine complexity, was even harder. After much reflection, we decided to go for it. I signed the order, and posted it, along with the deposit cheque. We didn't foresee how fast global warming would progress.

At COP23 in Germany, Donald Trump's withdrawal of the USA from the Paris Agreement a few months before hindered progress. Average monthly carbon dioxide levels recorded at Mauna Loa Observatory in Hawaii, increased to 405.31 ppm.

We were feeling the climate crisis through extreme weather events, like the frost that hit us that year. We read that trees provided a buffer against frost, so Sean nurtured native trees in the vineyard. It was good for biodiversity too.

Sometimes, people are confused about the meaning of the words *biologique*, biodynamic, and biodiversity. *Biologique* means organic in French. It is a term that can only be used for certified organic products. Biodynamic is a way of farming that is 'organic plus'. A farm must already be certified organic to become biodynamic. Biodiversity is the variety of plant and animal life in a particular habitat. Biodiversity, unlike the other two terms, is not a certification. It is a valuable component of a healthy farm and something that can be encouraged in all farms and gardens. Biodiversity in the vineyard, goes beyond low-level plants and grasses, to include trees, and hedgerows, and beyond insects, to include birds and mammals.

When we changed to handpicking all our grapes, and were therefore no longer obliged to keep the trellis clear for the harvest machine, Sean encouraged ivy to grow up trellis poles every 20 metres or so, to feed pollinators. Ivy flowers in late autumn, providing much needed pollen for the bees and their ilk, and it is a massive biodiversity booster. An oak tree with ivy, hosts double the number of species than without ivy.

Biodiversity is amazing. For years, we wondered why we didn't have a problem with grape worm. The name gives away why it is a problem for wine farmers. Bats are our allies. Bats love to eat the European grapevine moth that lays the worm eggs on the grapes, thus solving the grape worm issue for us. But bats will only fly about 30 metres into a monoculture with no relief. To go further, they need navigation points. Ivy and trees provide that. By encouraging ivy and trees for other reasons, we landed up solving a major headache.

Bats also need habitats, places to live, in the form of hedgerows, forests, cliffs with caves; which we are fortunate to have. I love the trees, and biodiversity, on our farm. When I walk along the hedgerows with guests, I introduce trees, bushes, and plants, by name. Oak, walnut, willow, fig, elder, hazel, hawthorn, blackthorn, dog rose, bramble, fern, nettle, horsetail. It's a growing list thanks to increasing biodiversity and my increased awareness. The vineyard floor is also a mass of biodiversity: clovers, grasses, vetch, wild orchids, buttercups and more. Each plant brings a benefit, for example, clover and vetch fix nitrogen. They draw nitrogen gas from the air and convert it into forms of nitrogen that can be used by plants, acting as a natural fertiliser, or 'green manure', as we call it in farming.

I explain how to pick up the differences in the vines and their characters: the old vine Merlot, the young, fiery Cabernet Sauvignon, the disease-resistant, old vine Semillon, the zesty Sauvignon Blanc. Respect is built by us knowing them; by getting close.

There are two perfect shady spots where I stop to talk about biodiversity on the tours we offer. The oak trees there seem to know me. When I touch their leaves, it feels different from how it feels with other oaks. We have formed a relationship. They 'see' me, 'hear' my voice, and sense me gently touching their leaves regularly. In the magnificent book, 'Braiding Sweetgrass', Robin Wall Kimmerer shares the indigenous American wisdom that if we see the earth, rocks, water, insects, plants and trees, as beings we respect them. In the magical book '32 Words for Field', Manchán Magan makes a similar point about the Irish language offering a sense of respect, and richness of nature. Our words change our hearts.

We had signed the contract for our kitchen renovation before the frost. If we had known it was coming, we would not have. Life had a way of making decisions for you. With the volume of vegetables and fruit Sean was growing, we needed space to prepare and preserve the bounty. The renovation would provide it. We moved into the Lodge accommodation so the renovation could start. The previous winter, I enjoyed hours planning the new kitchen. The walls that divided the galley kitchen, office, and shocking toilet, would be removed to make a large open plan space. Low ceilings would be removed to expose magnificent oak beams and stone walls dating to the 15th century would be uncovered.

It was audacious; the culmination of many years of planning, and organising. First, we repaired the roof and planned how to expose the beams and open up the room. Now we were down to the nuts and bolts of dividing it and fitting out the kitchen. I wanted solid wood furniture. I had seen too many medium-density fibre board (MDF) kitchens disintegrate in ten years. This one was going to last a lifetime.

The weekend before the renovation started, Sophia and Ellie took their paints and went crazy Andy Warhol style, on the walls. We took photos. Then it was gone. The new space emerged like a caterpillar turning into a butterfly. Natural light illuminated the exposed limestone – ancient compressed seabed – that had been quarried centuries before on our farm.

The finishing touch was a wood-burning cooker from Italy. Sean left for a two-week trip to celebrate his sister's 50th birthday the day the cooker disappeared. How did a massive chunk of metal get lost? The supplier had no answer. We wanted to be back 'home' for Christmas, but we could not move in until the cooker installation was complete.

The day before Sean's return, the cooker was found and installed. I kept up the story that it was lost, and we couldn't move 'home'. Sean walked in, and there it was, studded with red tiles that matched the terracotta of the floor, beneath honey-coloured oak beams and cocooned in cream limestone of our cosy 'new' kitchen.

Our dream kitchen had everything we needed for better self-sufficiency. Moving back brought a sense of settlement and peace. On Christmas Day, Sean started feeling sick, he had brought the flu back with him. He holed up in bed to care for his misery, then passed it onto me. I went down hard, my head crushed like it was in an iron clamp. Sean brought me tea and paracetamol. I lay in bed oozing infection, my eyes gummed up, my nose and sinuses blocked, my chest tight. We had planned to go to Bordeaux for the day, but I wasn't getting out of bed. Sean, Sophia, and Ellie decided to go. Before leaving, Sean came up to check on me. I saw pity in his eyes.

He brought me a glass of water, and they left. I felt unloved and extremely unwell. It felt like the ultimate rejection, proof that our relationship was dead. When they got back late afternoon they found me where they left me. Sean brought me a honey lemon.

Days later I woke to the sound of an owl hooting, and then a robin singing. I was out of the prison that had held me so tight; left me so dark. What was this magic that gave energy to move, think, imagine, dance; rather than sit and cry? Looking back at the year, I realised that I was getting sick more than before. Something was not right. I read that a bad relationship could affect your health and your immunity. Stress was also a major factor.

Francois, the friend that Sophia and I tubed with in St Cirq la Popie, emailed to say he and his family wouldn't be visiting that summer. They had been through a rough patch, and he and Jenny were getting divorced. Tears filled my eyes. It was the end of their family as we knew it. They were in the transition zone, moving from one era of life to another. Sophia and I had not been wrong about the coldness we felt on arrival in St Cirq. There was nothing simple about partner relationships, and longevity was no guarantee of ongoing success.

My brother, Garth, visited for two days. I hadn't seen him in fourteen years; and he hadn't met our daughters, except on screen. Garth lived in Canada; almost on the opposite side of the world and a nine-hour time difference, away. Like us, he had been building a business, and a family, and hadn't had time, or money, to travel. He was in Europe for a business conference, a good excuse to see us. I hadn't met his sons, either. That was about to change. I was heading to Canada for my mum and dad's 80th birthday celebration. Like Sean's trip for his sister's 50th, it would be a solo mission. We were doing more apart than together. On the face of it, it was for time, money, and to avoid carbon dioxide emissions, but it was also a sign of a growing divide.

I visited my sister, Foo, in Los Angeles en route to my parents. An upmarket restaurant in Santa Monica had organic beef and organic chicken on their menu; but not a single organic wine. An Italian restaurant in Hollywood had no organic wine either. I would not support non-organic wine. Following Foo's example, I turned to vodka.

When I touched down in Canada, I sought out organic wine to share with Mum. Dad preferred whisky. In a large wine shop with thousands of references, I found eight organics; none of which was the Pinot Noir variety I was looking for. Knowing what I knew about systemic pesticides in wine, I set aside my Pinot Noir desire and went for one that was certified organic.

We took walks, and chatted for hours over every meal, catching up, and reminiscing. It felt like borrowed time. Since their return from South Africa, Dad had been struggling with health problems. He had a wound that would not heal, and his cognition still was not right. A year of visiting specialists across Vancouver Island had yielded no satisfaction. He continued to fall and forget things.

'He seems invincible. Such a powerful character, you think he's never going to go,' said Foo as we ran together.

Shortly after, a specialist discovered that Dad had a blood disease called Polycythemia, a type of blood cancer, that could be treated with mild chemotherapy. He stopped falling, his cognition returned, and he got his driver's licence back. He cracked jokes on the phone and appeared healthier than he had been in years. Perhaps he was invincible.

I, on the other hand, was not. A few weeks after my return from Canada, I hit a wall. I couldn't drag myself out of bed. I felt nauseous, at times so sick, I had to crawl to the bathroom. A dark gloom fogged my brain. I was tired but I couldn't sleep.

CHAPTER 11

Broke back winery

I felt like lying in the bed and ignoring the world. I didn't even have the energy to read. Sean kept a wide berth, not sure what was going on with me. A close friend had been off work for months with burnout, so I knew the signs. They included fatigue, insomnia, sadness, irritability, and vulnerability to illnesses, symptoms I had experienced in recent times.

Sean encouraged me to see a doctor, but since we didn't have one, I had nowhere to turn. This was not something I could go to emergency for. Sean bought vitamin D drops and encouraged me to take them. Eventually, with spring growth and more sunlight, I felt some energy return. I forced myself out of bed. I was moving through mist instead of thick gray fog. Slowly, as if with the sun's increasing force, my strength returned.

An American couple, living in Belgium, joined me for a multiday walking tour, a needed push to get outside. At

the *Tour des Vents* restaurant, in Monbazillac, I described
botrytis noble rot, the fungus that concentrates the sugar
and creates sweet wines like Monbazillac, Saussignac,
and Sauternes, over aperitifs of potato and aioli, tiny
ham cheese sandwiches with parsley and chives, and
quail eggs. The north-facing slopes, and the river below
us, were part of the secret, aiding the development of
botrytis, particularly on the thin-skinned Semillon grapes.
In autumn, mist from the Dordogne River creeps up
the slopes in the early morning and then slowly drains
away. The misty mornings, followed by sunny afternoons,
create perfect conditions for the development of *Botrytis
Cinerea*, noble rot. It attacks the Semillon and makes
micro-perforations that allow the water to get out but do
not allow the bacteria to get in, a little like the pores of our
skin. The result is concentration of sugar, acids, minerals,
and flavours.

An *amuse bouche* of courgette mousse with tiny rolls
of courgette and fresh cheese, sprinkled with *fausse terre*,
arrived in a textured white bowl with wide curved edges
that demanded caressing. The *fausse terre*, false soil or
earth, was a savoury crumble of parmesan and squid
ink. The dark brown offset the geometrically patterned
spirals, pale green mousse, and white cheese, visually and
texturally.

Linda looked up after a spoonful. 'Wow, this is so good.
I never met a zucchini I liked before, but I love this.'

'Picked fresh, they can be almost as sweet as sweetcorn,'
I said.

'Amazing. I never knew,' she replied.

'They are rewarding and easy to grow,' I said. 'The taste
difference between just picked and one that has spent

weeks in a supply chain is dramatic. The bitterness that puts people off zucchini, or courgettes, as we call them over here, is from too long between picking and eating.'

I caressed the bowl before it was whipped away; its voluptuous curves reminded me of a Henry Moore sculpture.

'How are you enjoying living in Europe?' I asked.

'We love it,' said Linda. She paused as starters of fresh green beans blanched and finely chopped with smoked salmon and froth of bergamot were served. 'It was tough at the start. When we arrived, I got a terrible SAD, you know, Seasonal Affective Disorder. Winter in Belgium is gray and dark. I felt like I was in a fog. I had no energy; I didn't want to get up. I have always been a positive and energetic person but that disappeared. I felt like I was sick.'

'That sounds like what I experienced recently. I didn't realise SAD could be that bad.'

'For a long time, we didn't realise what it was. I didn't know what to do,' said Greg. 'Linda wasn't herself at all. We thought maybe it was homesickness.'

'It was so bad, I even considered going back to the US and leaving Greg to finish the three-year contract on his own. But I was sure there was more to it than homesickness,' said Linda. 'We read up and found that SAD matched how I had been feeling. It also got way worse in winter. That was another tip-off.'

'What did you do to fix it?' I asked.

'Greg bought me an amazing lamp,' said Linda. 'It changed my life.'

'It's a simple solution,' said Greg. 'But it works.'

'Truly,' said Linda. 'I'll send you the details if you like.'

'I'd love that,' I said feeling like she had offered me a magic wand.

Our mains were presented at the identical moment, like a choreographed ballet. They were a delight for the eyes, served on rectangular white plates decorated with artistic sweeps of walnut extract, edible flowers, and baby pea shoots. The main features, veal fillet, a polenta circle stuffed with roasted walnut chunks in a tube of Echourgnac, a local cheese rubbed in walnut liquor, and mini cannelloni of spinach, were artfully arranged.

'There's no chance of SAD, with food like this, and weather like that,' said Linda, pointing to the sun-drenched vineyards draped over the hillsides below us.

'I feel relieved after hearing your story. It may be what's been happening to me in winter,' I said.

'SAD is not to be messed with,' replied Linda. 'It's serious and it can be fixed. It's like menopause. No one talks about it, but we need to.'

'You're so right,' I said. 'I talked about menopause in my last book. I thought it would pick up media attention, but it didn't. It's a topic people don't like to talk about. Perhaps menopause and SAD reinforce each other.'

'You could be onto something,' said Linda.

'I'm staying out of this,' said Greg. 'What's for dessert?'

Perfectly timed, fresh *gariguette* strawberries, strawberry sorbet, pistachio mousse, and dark chocolate ice cream with touches of mint leaf and mini *batonnets* of shortbread, were served on blue plates decorated with coulis of strawberry. It was another work of art, so beautiful I didn't want to destroy it. But temptation got the better of me, and I leaned in for a spoon, swiftly followed by another. The well-being of walking, great food, and the gorgeous day, enveloped me. My dark days, a few short weeks before, were like a bad dream.

That weekend I stumbled to the landing to pick up my phone, wondering who was calling so early in the morning. The phone flashed 'Dono', a good friend of Russ, Steph, and mine.

'Caro,' he said, his voice choking. 'Russ is dead.'

'No.' Tears rushed down my face and I sank onto the top step. I dropped my head between my legs, feeling dizzy.

'I can't believe it,' said Dono. 'Our best friend, my brother; he's gone.'

I felt gut punched.

'How?' I asked, my voice rasping.

'A brain bleed yesterday. He died quickly. The doctors say it was painless. They say he would have been dead before he realised what was happening.'

We sobbed, then sat with our grief for a while. When we regained control of our voices, we talked about our friend, his wide smile, and generous spirit. As teenagers, Russ was always ready for fun, dancing, and laughing, in his violet lab coat, the life and soul of the party. He was a lover of music and people, an adventurous man who would drop everything and go the extra mile for family and friends. I felt like he was still with us; that this was a mistake, that someone would call and say he was okay.

Dono was like Steph and Russ to me; even though we didn't see each other often, we were close. We knew each other so well from shared adolescence, from sharing the moment in our lives when we were transitioning from kids to adults. There was something soldering about passing through mighty transitions like that with someone; you were closer for it. Like what Sean and I were going through now, as we transitioned through menopause; through the straits we were currently navigating, and into the empty nest. Some relationships would break in the transition, others would come out stronger.

That Russ was dead seemed impossible. I could hear his voice in my head, our promise to meet in Italy.

After hanging up, I made tea and went out to the office. Steph called as I got there. I stood at the glass door of the tasting room, tears flowing again. We reminisced, then said a prayer of love and healing for Russ' family, and his safe passage onwards. He died soon after turning 49.

As we went from spring, into summer, I felt grief for Russ well up at unexpected moments. As I rinsed glasses in the tasting room, I remembered our exchange about how he loved to buy wine from estates he had visited. Walking down into the valley between Saussignac and Gageac with a walking tour group, I felt a flash of how much he would have loved exploring this place, and a wave of sadness that he would never see it. When I picked up my guitar, I thought of him.

The rising tide of tourists forced me to get on with it. Multiday tours, particularly walking tours, were growing in popularity. I wanted to increase the active tourism part of our business. I loved walking tours; they kept me fit and offered the opportunity to see our local landscapes up close. Favourites included our commune of Saussignac, St Emilion, and nearby Monbazillac, where I had been with Linda and Greg.

Lee, a school friend who lived in Australia, took a year out to do her MBA in Copenhagen. She bussed down to see us for a few days. We ate cherries, walked, and chatted by the pool. Her visit forced me to take downtime before the peak season kicked in. It was grounding to hang with my old school friend, to share views on life, and how it was evolving. I needed to take breaks. I couldn't expect my body to keep the same rhythm as it had in my thirties when we started our vineyard adventure.

In the heat of summer, I returned to Monbazillac with another small group. We started our day with a walk around the fairy tale castle with four turrets, majestic on the valley ridge above Bergerac town. From there our walking route took us into vineyards then into light forest where we were thankful for the shade. The temperature was predicted to hit 30 degrees Celsius.

By the time we reached *La Tour des Vents*, sweat poured down our faces. Refreshed by jugs of cold water we chatted and soaked up the views of Bergerac and the surrounding patchwork of farms. The restaurant pitched the *amuse bouche* of cold watermelon and goat's cheese soup, and the starter of cold beetroot mousse with beetroot chips and a touch of savoury cream, perfectly for the day.

Leaving my guests with chilled white wine I stepped outside to check the forecast for the afternoon. It had edged up to 33 degrees. Too hot for guests to walk back, but okay for me. Sean was on call in case we needed help, as always with walking tours, but I didn't want to bother him. In the washroom, I flicked cold water onto my face and stroked it down my neck, arms, and legs. I'd be okay.

The vegetarian main course of red pepper and yellow courgette, stuffed with potato and herbs, was light, fresh, and seasoned to perfection. As we finished our *Tarte aux fruits rouges* with chocolate ice cream and mint sorbet, I recommended that, given the heat, I leave them to enjoy the views over coffee and more water while I walked back to fetch the car. They readily agreed. I ordered another jug of water, and coffee, then settled the bill and strapped on my backpack.

The start was shaded, and I felt fine. When the trees ended, the sun sizzled. My scarf, soaked in cold water before I left, and hung over my shoulders, quickly became paper dry. Sweat trickled from under my hat, down my forehead and into my eyes. I took the edge of the scarf and wiped it away. I kept striding. A star floated across my eyeball, then another. I felt dizzy.

I was about halfway when I realised that I wouldn't make it on foot. I needed to get a lift. I turned out of the vineyard and reached the tar edge, feeling wobbly. I stuck my thumb out. A smart new van flew past. Feeling faint, I backed up and sat down on the grassy curb. Sweat streamed down my face. More stars bounced in my vision.

Another car appeared in the distance, I stuck my thumb out. The old white Clio pulled over. I felt a wave of relief, struggled to my feet, and fell gratefully into the passenger seat. I thanked the driver for saving me from the heat. The young man said it was nothing. Minutes later he dropped me in the shady parking lot. I felt a welling of love for him and humanity as if his act of kindness had connected me not only to him, but to the entire universe.

I should have asked for his contact details to send him more than my verbal thank you, but I couldn't think beyond shade, water, and not breaking down. In the castle washroom, I doused myself in water and drank handfuls, then sat in the shade to gather myself. In the car, I set the air conditioning to 23 degrees Celsius, ten below the temperature outside. I waited for it to take effect, then drove back to collect my clients, looking as if nothing had happened.

The experience delivered two messages. My body would not accept the abuse that it accepted in the past, and global warming was no joke. I always checked the forecast before a walking tour and planned for the conditions. This time the forecast was under what the real temperature turned out to be. The climate crisis was sending us unexpected extremes. I knew to be wary in the future.

Sean, Sophia, and Ellie went to the Netherlands for a holiday, first to Amsterdam for a couple of days, then to visit Ad and Lijda. I could not be away at such a critical part of our tourist season. We had a full house of resident guests, and tour guests booked for the week concerned. It was too much to expect Ros, the dynamic, young woman that worked with me that summer, to cope on her own. Sean and I were not doing anything together. Like continents drifting apart, unperceived slow separation was becoming a gaping ocean.

Harvest came fast on the back of a hectic tourist season. Ros decided to do a year-long part-time wine tourism diploma. It meant she could join us as an apprentice for the year. It was a lucky break for our business and for my mental and physical health. But as harvest drew to an end, Sean reached out to swing a vat's floating lid into place. Something snapped, and he was in such pain he struggled to get back down the ladder. He hobbled across the courtyard and took to the nearest bed. He didn't know what had happened. In the 'medical desert' of rural France, our chances of finding out from a qualified doctor were slim.

Doctor Internet advised that Sean needed potentially paralysing surgery, or, if he were lucky, rest and anti-inflammatories, would solve it. Naturally suspicious of 'rest' I wanted to do something. For me, action was a solution to everything. I needed to revisit the lessons I learned at the conference in Alsace about less action and more thought and heart. Sean rightly opted for rest. He stayed in bed, immobile, crawling out in agony, to go to the toilet, a couple of metres away. I dug through the medicine box for all the anti-inflammatories we had.

My first worry was Sean, my second was how we would do the winery work. For Sophia and Ellie, the second quickly became, 'Who will feed us proper food?' While our daughters faced the horror of lentils and rice every day, I faced becoming an unwilling cellar rat, back in the winery, as I was for our first harvest twelve years before, when Sean chopped his finger off.

Harvest was over, but winery work was still in full swing. The white and rosé wines had finished alcoholic fermentation and needed racking. 'Racking' is taking the clear wine off the sediment, something winemakers do at key points in the wine's life cycle, like after fermentation. The red wines were close to finishing their fermentation but still needed pumping over or punching down twice a day. We agreed that the winery could be left for 24 hours with no harm. Hopefully, Sean would be back to normal by then.

The following day, Sean was still immobile. We couldn't leave the red wine any longer. I had to tackle the pump-overs. I didn't know where anything was, or how the work was organised. Sean dragged himself out of bed to guide me. Crossing the courtyard, he looked like a

90-year-old, barely able to shuffle, taking one tiny step at a time. I felt sick with worry. We muddled along, Sean directing as patiently as he could, given his extreme pain, and me trying to take direction graciously.

On the third day of no improvement, I insisted we get him to a doctor and ramped up my search. I called every doctor within a one-hour drive of us. There were about thirty of them, and they all said a firm '*non*'. They were full to bursting and unable to take more patients. I pleaded and grovelled, but it made no difference. Eventually, a secretary in a nearby village said they were hoping for a new doctor in January, and that I should call back then. I set a reminder in the diary.

Until we experienced Sean's back incident, I thought the rural France 'medical desert' story was exaggerated. When we arrived in Saussignac, there was a doctor living in our village with his consulting room in Gardonne, five minutes away. He made house calls to the previous owners. At the time of writing, there was no doctor in Gardonne and our retired doctor in Sigoulès, ten minutes away, had not been replaced.

After failing to find a doctor, I called the emergency phone service, *le 15, SAMU*, which offered 24*7 access to a doctor's advice. I explained that I was calling because we did not have a '*médecin traitant*'. In France everyone is attached to a GP, their allocated '*médecin traitant*', the doctor who is called in an emergency and who shou[ld] know your medical history. After ensuring he unders[tood] the medical desert problem, I described Sean's sit[uation] how it happened, where and what the pain was[,] immobile he was.

'If you have no doctor, you need to go to Emergency,' said the specialist.

I explained that local Emergency, while it was very good, was always an extremely long wait for anything that was not an immediate threat to life.

'Sean isn't in a state to go anywhere, especially not to sit in a waiting room for hours,' I said.

'Okay, you probably don't need to go to Emergency. Given your description, rest and anti-inflammatories should fix it,' he said, confirming Doctor Internet's opinion.

We were already doing both of these things. I thanked him and hung up. I would have to reconcile myself with 'rest' for Sean and 'cellar rat' for me. We had to rack the white wines and the rosé wine. They could not wait another day.

Sean instructed, leaning in agony, against the winery wall, as Ellie and I took on the delicate operation. Each wine needed to be pumped out of its fermentation vat and into a new maturation vat, using a u-shaped pipe-end. The aim was to take the wine and the fine lees, but leave the sediment and heavy lees behind. In French, this task is called 'soutirage', to 'pull from below', as the sucking end

pipe faces upwards.

fine lees, not the heavy lees, so the stop was critical. When we wanted the lovely pea-green live-drab ones lower down. I g the pipe in position at the ed the pump. Sean watched the n to stop. It was a three-person mplish all this work on his own?

On the Wine Spirit Education Trust courses we offered, I had managers from two organic St Emilion Grand Cru estates express awe at what Sean achieved, operating our vineyard and winery, and creating our wide range of organic, biodynamic, and natural wines, on his own.

We were both stretched in the season. No wonder our relationship was taking strain. Since Sean rejected the book I wanted him to read, I had ignored our rift. Forced to work together in the winery, and feeling compassion for his painful state, I knew I had to do something soon. For both our sakes, we couldn't continue as we were. Like the wines, we needed to move onto the next phase, whether that was fixing our rocky relationship, or taking separate ways.

The wines had finished their alcoholic fermentation. The racking work moved them to vats where they would pass their maturation phase in a restful state. With the wine moved, the next step was to fit the floating lids, the action that had caused Sean's back problem. I followed Sean's instructions, but I couldn't do it on my own or with Ellie's help. The lids were too heavy and unwieldy. I called Thierry Daulhiac, our friend who had saved us many times, starting with our first Saussignac dessert wine in my book 'Grape Expectations', then picking up the ball on selling our grape skins, and visiting California, in my book 'Saving Our Skins'.

Within an hour, Thierry was with us. Sean gave guidelines on what he needed. It was second nature for Thierry; he knew exactly what to do. Like Sean, he knew this work so well he could have done it blindfolded. He climbed the ladder and moved one lid, fitted it, then swung over, and fetched another, to do the same. I held my breath; it was like watching *Cirque du Soleil*. After fitting the second, he moved the ladder to another vat and started the winery ballet again. Despite our investments over the years, it was an eccentric winery, perfect for a trapeze artist.

With the whites and rosés stored in maturation vats, our thoughts turned to the reds. After tasting them, we decided they were tannic and full-bodied enough. They could not wait for Sean to recover. The reds usually took a night to run off the 'free run wine', the wine that runs freely out of the tank by gravity. That was followed by a day of intense physical work, digging the grapes from vat to press to release the 'press wine', then moving them from the press to the compost heap.

I had back-to-back tours, and no free days coming up. Alone, I wouldn't be up to the physical demands of the job. We could not ask Thierry, or another winegrower friend, to stand in for such a long job. They were up to their necks in winemaking too. None of them would have a day available. I took off for St Emilion, half-listening to the radio, and half-grappling with what to do about draining the reds.

My mind was still gnawing on the problem, as I took a case of wine from the boot, to deliver to ETS Martin, one of my favourite wine shops in St Emilion. A black crow swooped down. I looked up, and stumbled on the cobblestones, caught myself, and saved the case. My mind was elsewhere. Starting my day in a pool of red wine on *Place du Clocher*, was not what I needed.

A couple of hours later, as my group tasted with Guy Petrus Lignac, owner of Château Guadet, I stepped outside to call Thierry, to see if he knew anyone who could help us for the pressing day. The Guadet garden was a haven of peace between the original protective city wall of St Emilion and the house which fronted onto *Rue Guadet*. The Lignacs had been there since 1844 and had been growing vines in the region for a lot longer. In the garden, tomatoes, basil, and roses cosied up to one another, trees provided shade, and the whole was serenaded by bird song. It felt miles away from the busy main street outside.

Vince, son of Guy Petrus, strode out of the house, on his way to the winery. Wiry and lean, with shoulder-length dark, wavy hair, and an easy-going manner, his face lit up with a wide smile seeing me at the outdoor table. We exchanged *bises*. I launched into telling him about Sean's plight, and our thorny problem of how to press the reds. I hadn't thought of asking Vince if he had suggestions, or could offer help. The story poured out of me unbidden; like the universe was guiding me.

'We're in a quiet week, I could manage without Gildas, my apprentice, the day after tomorrow. He could come and help you,' said Vince.

'Thanks, Vince. Are you sure?'

'Absolutely. I'll check with him, but I'm fairly certain he'll be okay with it.'

'You've made my day. Thank you.'

'My pleasure, it's nothing,' he replied.

At home that evening, a weight lifted off Sean's sore body, when I told him about our small miracle. I set up a one-day seasonal contract for Gildas, and two days later, the fit 20-year-old was at Château Feely, digging out and pressing the reds with Sean's guidance, as I led a group day tour in the tasting room. Late afternoon, the press wines were in the vats, the skins were on the compost, and the winery, vats and press, were spotless. We thanked him heartily, and I sent a thank you to Vince. With the reds resting peacefully in their maturation tanks, Sean could settle into some rest too.

On the seventh day, Sean was able to move without extreme pain. By the tenth day, he was back on his feet. It turned out, that rest, in this case, was the solution. We needed to respect messages from our bodies and give ourselves the right to rest before it became a crisis.

The previous time Sean put his back out, we were living in Dublin. We had recently laid out our dream to go wine farming in France, in the hope that the universe would help us achieve the unachievable. While I was out playing golf with a friend, Sean bent down to pick up a pillow and found himself seized up on the floor in agony. When I got home, he was lying on the sofa, unable to move. At first, I didn't believe him. I thought it was a ruse to get out of the housework he had promised to do that day. That time he needed a couple of days rest. This time it was ten.

As Sean got back into normal life, I was hit with another bad upset stomach. I realised that each time I had been sick with cramps and nausea I had eaten peanut sauce. Somehow, I had become allergic to peanuts. My sister Foo had developed a similar allergy to seabass but only farmed seabass, not wild seabass. Something in the farming of seabass, perhaps a fungicide or an antibiotic, was the cause. Like me, it took her three bouts to realise the common denominator. The relief was divine. It was like when I discovered SAD from Greg and Linda like a menhir had been taken off my back. I would not touch peanuts again.

But across the globe, we continued to be addicted to manufactured energy, particularly, to carbon dioxide emitting forms of energy. It was not as easy as saying no to peanuts. COP24 in Poland was addressed by fifteen-year-old Greta Thunberg, who, through her School Climate Strike movement, Fridays for Future, and no-holds-barred speeches, had raised more climate crisis awareness in a few months than the COPs had in 24 years. That November, the monthly average carbon dioxide in the atmosphere, recorded at Mauna Loa Observatory, increased by close to three, to 408.21 ppm.

Our kitchen renovation transformed our home life. We had room to preserve and cook food, and to be together as a family. The space between the cooker and the dining table was big enough to be an impromptu dance floor. Sophia, Ellie, and I, took to dancing after dinner, swirling and whirling, feeling the pure joy of moving to music, letting everything go.

Sean's back was healed, and my gastro upsets were over, but we were still operating like a business partnership. At home, we were too sucked into the business to see ourselves for who we were. We needed a reboot. We booked a trip to Ireland, for marketing, to see friends, and to keep our connections with a place that felt like home for us. It would offer needed distance from our everyday life.

CHAPTER 12

Showdown in Dublin

Before we left for Dublin, people took to the streets across France rioting, looting, and burning cars. Chaos and random acts of violence exploded over a planned carbon tax that would punish the poor rural population the most. The money from the tax was not specifically earmarked for environmental action, instead, it would go to general government coffers. The *Gillet Jaunes* – Yellow Vests – movement, gathered momentum causing embarrassment for President Macron at the G20 conference in Argentina. While a carbon tax was a good idea, given the climate crisis, the way it was proposed, and its handling, inflamed French society.

As an economist by training, I kept in touch with Economics. An English economist, Kate Raworth, had created a new way of economic thinking she called 'Doughnut Economics'. It assessed the resources of our planet compared to our consumption. She posited that we needed to cut back to live inside the donut, formed by the borders of what our planet could support. She created a circle within a circle, a doughnut, showing all the key parts of our economy.

The doughnut consists of a central social foundation that needs to be solid. This includes meeting the needs of water, food, health, education, income and work, social justice, peace and justice, political voice, social equity, gender equality, housing, networks, and energy. When these are met within the ecological capacity of the earth, we are in a happy place, the 'safe and just space for humanity' within a 'regenerative and distributive economy'. Outside the doughnut, if we overshoot the carrying capacity of ecological systems, we create climate change, ocean acidification, chemical pollution, nitrogen and phosphorus loading, freshwater withdrawals, land conversion, biodiversity loss, air pollution and ozone layer depletion. Reading her theory, it was clear we should all be taking to the streets on climate strikes with Greta Thunberg.

Raworth was one of the few commentators motivating for reduction of consumption to within the boundaries of our planet. The ecological discussion was often hijacked by green businesses, proposing business as usual with some green tweaks, still pushing for ever-expanding growth.

Changing our perspective from the consumptive capitalist model to a well-being and contentment model does not happen overnight. Changing expectations set by what we see online and on television into realistic expectations for ourselves and the long-term health of our planet is going to take a mind shift.

I decided the best place to start was with myself, with small steps that would add up. I requested handmade gifts from my family rather than bought and thought of Khulu in Zinkwazi. I changed to locally made solid bars of soap and shampoo instead of liquids in plastic bottles.

On Christmas Eve, we gathered at the table in our 'new' kitchen. The wood-burning cooker brought a warm glow to the room and we shared stories. Sean had started a tradition of each of us writing a story to share on Christmas Eve. It has become a loved part of Feely family Christmases. Sharing and listening to the stories was better than glitzy gifts. I was overwhelmed with gratitude for our home and our family.

Two days later the feeling of well-being disappeared soon after we touched down in Dublin. Sean didn't like the Airbnb I had booked for us despite its perfect position walking distance from our friends and Dun Laoghaire pier. We couldn't agree on what to do. Sophia and Ellie wanted to go shopping. I wanted to go walking. Everything was edgy and didn't feel right. I felt like we weren't living a life in tune with our ethos if we drove everywhere and spent our lives shopping. But our daughters were teenagers and wanted to do exactly that given we were in a big city at last.

Sean wanted us to do things as a family, the four of us, regardless of what it was, and he wanted to do what the girls wanted to do. I believed he prioritised them over me. It was this issue that the book I wanted him to read had said was a marriage breaker. Feeling frustrated, I slammed the door and went out for a walk. It was not a mature, adult response. It was a spontaneous reaction to what had been building for years. When I returned, I felt calmed by the sea, like I was in a better place to cope.

The following day, our 21st wedding anniversary, Sean gave me a card that said he was looking forward to the next 21 and I cracked inside. I didn't want another single one if it was going to be like the last few years. I was discontented with where we were in our relationship. We were like an estranged brother and sister sharing a house. I kept sounding the alarm that things were not right, but Sean ignored me. We went through the motions of our days but there was no spark, no fun, no joy in each other. When we lived in Dublin, we shared romantic dinners every anniversary. Now it was a random card. I didn't give Sean anything. I didn't feel like our relationship warranted celebration as it was.

Sick in my belly and with tears bowling in my eyes, I went to the bathroom to control myself. I couldn't voice what was in my heart with our daughters in the apartment. I discreetly called my friend Aideen. I needed to talk to someone who was relationship-wise before I did something I would regret. She came over. We took the bridge over the train track to the sea.

As we stepped it out on the Dun Laoghaire pier, I unpacked the backstory starting with the bursting point of the card that morning.

'Our relationship's so broken Aideen. Sean and I are no better than business partners. At home, we barely talk except about work. Sean sinks into television every night to avoid reality. We've lost our spark. Our romantic relationship is dead. But we're so deeply entwined in our business, and as a family, that I feel like there's no way out.

Sean's oblivious. He thinks nothing's wrong. What am I going to do?'

Aideen waited a moment before she replied.

'As you say, Sean doesn't realise. The first thing to do is to organise a proper heart-to-heart. Lay out how you're feeling really clearly, and without blame. Ask to go for a walk so that you can talk through everything openly. Say how the current situation makes you feel.'

'Terrible,' I said.

'Okay,' said Aideen. 'You can say that, but not in a blaming way. Find a way to say it, that makes it nobody's fault, just how it is.'

'But how do we fix it?' I asked.

'By talking. You can't dictate the solution. You can suggest potential steps, but don't be prescriptive.'

'I hate that,' I said. 'I want it sorted out now. Concrete steps.'

'I know, but that will make it worse. Sean will feel cornered,' said Aideen. 'Communication is key to resolving everything.'

'You are so wise,' I said. 'Tell me more.'

'The most important thing is to be kind to each other. Think before you speak. There's a concept we use in communication training. It's called 'beware the four horsemen'. The four horsemen are criticism, contempt, defensiveness, and stonewalling. They're bad communication patterns, for any relationship. Once you know about them, you can be more aware, and notice when you fall into them. You can think about how to change your approach.'

'Hearing you say that I can see straight away that I criticise and go defensive, and Sean stonewalls. It doesn't do us any good.'

'Being aware will make a difference,' said Aideen.

'It's not only that though,' I said. 'I feel like we don't have fun together anymore. We never do anything just the two of us.'

'It's easy to fix that, organise some date nights. Do something you both like to do. It's tough working together and probably made worse by the stress of farming.'

'Yes. When it's the two of us, we inevitably talk about work,' I said.

'And understandably. You've worked hard, and you should be proud of what you've created. Look at what you started with, and where it is now. You need to celebrate your achievements too.'

There was much to celebrate and be thankful for, but I didn't feel like that. I felt like my world was caving in.

'Remember also that we're complex beings. You can't get all your needs met through one person. That's why friends, family, groups that 'get' you, beyond your significant other, are important.'

Aideen was onto a delicate subject, we lived far from our extended families and old friends, but even our near friends had taken a recent hit.

'You're so right. A part of the community we had in Saussignac has left. Laurence moved back to *Pays Basque*.'

Aideen knew Laurence, my friend, and Pierre, her husband, who had been our bottler for many years. Aideen had touched a sore point.

'Then it's time to make a new friend in the village.'

'It's tough to do when you have so little time.'

'I know,' said Aideen. 'I'm so sorry Laurence has left.'

We reached the end of the quay and sat on the seawall looking out to sea.

'The most important community is my husband and that's where I need to start. You've given me some good tips, but I feel like Sean's going to ignore them. How do we work this thing out?'

'Look into his soul and see what's going on. Literally, look into each other's eyes. You'll be able to see more deeply what's going on, sadness, regrets, fears, excitement, love, possibility.'

'It can be hard to do,' I said.

'I know,' said Aideen.

We sat for a few minutes, listening to the sea lapping on the wall.

'The way we show love can be different for different people. For some it's words of affirmation, for others quality time, or perhaps receiving gifts, or acts of service, for others it's physical touch,' added Aideen.

'I never thought about that. I need physical touch and affirmation, whereas Sean doesn't need them.'

'There you go. What does Sean prefer?'

'Quality time and acts of service. Things I haven't been giving with the pressured growth of our business,' I said.

'Exactly,' said Aideen.

We got up and started to walk back. Gulls lifted off the rocks and rose out towards the open sea. On the opposite side, the harbour water was quiet and dark.

'Little things can go a long way: a picnic, a note of appreciation when you're away, a gift, dinner together with no kid talk and no farm talk. You'll be surprised at how small shifts can make a difference.'

I felt like Aideen was giving me tools to find a way back into our relationship. She was a guru.

'Another good one is to imagine your future together. How do you want to be with each other in ten or twenty years?'

'Sean hates doing that.'

'Okay. So, take another tack, how about looking through your photos of shared memories together? Appreciate how far you have come and grown together. Approach imagining the future subtly by using your shared memories as a starting point. Then imagine future memories that will be your photo album of the future.'

'Thanks, Aideen,' I said, turning to look into her eyes, tears in mine. 'You're amazing.'

She held my gaze.

'A long-term marriage is worth fighting for. The depth of time can't be reproduced. Those memories, maturity, a knowing. Like an aged wine... You know what I mean.'

We laughed. I pulled a tissue out of my pocket to wipe my tears.

The next morning Sean and I opened our hearts as we stepped it out on the same Dun Laoghaire pier. I didn't know how he would react. Part of me expected him to get angry as he did with the book suggestion, but I was too unhappy to let that stop me. I remembered Aideen's wise words. As I spoke, I found myself rewording, stopping myself from stepping into criticism.

'I'm so unhappy SF,' I said using the nickname I gave Sean soon after we met. 'I don't want another 21 years of marriage if it's going to be like the last few years. When I read your card, it brought it all bubbling up for me. It's neither of our faults. It's lack of attention from being caught up in building a business and raising a family. But it isn't the relationship I want.'

Sean turned to me.

'I feel like all you do is criticise me,' he said.

'I don't mean to do it, but I can see I have in the past,' I replied. 'I'm going to watch myself for that. I feel like you shut me out.'

'It's my way of coping when you criticise. Do you think I want to be like that?' he asked.

'No. I know you don't. It's communication patterns we have slipped into. Things that don't help us, or our relationship. It's not only our communication. It's also about spending time together, just the two of us. I know we have a lot of time together as business partners and parents, but we don't give any time to our romantic relationship. We can't be happy if we don't give that as much attention as the other parts of our lives.'

As I used the communication lessons Aideen had shared, I found Sean listening instead of becoming angry. I realised that part of what made Sean react the way he did, was my way of communicating.

'How do we do that?' asked Sean.

'It can be as simple as going for walks together.'

'Or gardening together?' said Sean.

'Maybe. As long as you don't boss me.'

We both laughed and it dissipated the tension. Whenever I was in the garden, Sean bossed me around since it was his domain, like in the winery. I didn't take it well. When we lived in Dublin, gardening was our hobby and I loved it. Now I rarely did it, partly due to time, partly due to seeing it as part of the farm business, and partly due to this subtle power issue.

Sean was hurting as much as I was. We began to unpack and recognise where things were going wrong. Building a business and raising a family created pressure on our relationship. That pressure needed to be recognised and dealt with, not ignored. We had work to do to repair our trust, but we both wanted to do it.

'I love you, Caro.'

We looked into each other's tear-filled eyes and knew we had something too good to throw away. I reached for his hand, rough from constant outdoor work, and squeezed it tight.

'I love you, SF.'

We held hands as we walked the last section of the pier. By the time we got back to the apartment, the energy between us was transformed. There was mutual recognition of each other, of the love we shared. We had thrashed out where we needed to beware, and what we could do to get our relationship back into a better place. I felt light and happy, totally metamorphosed from how I felt before we set off.

If I had taken on that discussion without Aideen's wise words to guide me, it could have ended in a different outcome. I was shocked at how close I had come to blasting out. Long-term relationships need care and constant repair. As Shakespeare wisely said, 'The course of true love never did run smooth.' On that walk in Dublin, we committed to the reconstruction.

Soon after our return, I discovered a book that built on Aideen's advice, Thich Nhat Hahn's 'True Love: A Practice for Awakening the Heart'. Thich Nhat Hahn or 'Thay' to his followers, was a Zen Buddhist monk, poet, writer, teacher, peace activist, and founder of

Village, the largest Buddhist Monastery in Europe, amazingly, a mere ten-minute drive from Feely farm.

'Understanding is the essence of love,' was his key message. Understanding your partner's deepest troubles, aspirations, and suffering, is necessary to truly love them. This takes time and 'looking deeply'. I could see Aideen nodding sagely at his advice. In this small wonder of a book, he recommends several mindful breathing exercises and simple mantras including 'Dear one, I know you are there, and it makes me happy.' He had a gentle way of saying things that could bring a chuckle to important and difficult messages.

Back in Saussignac, we began the slow rebuild and I turned to Aideen and Thay's advice to help me be a better partner. Unfortunately, that didn't extend to cooking. My lack of *gourmandise* in food preparation had always been a bone of contention. Sean took over all the cooking. I organised a night away when we had a friend to stay that could look after our daughters and the menagerie. Sean started taking walks with me on Sundays. I reached for his hand when we walked, and he held it instead of letting go.

Sean was preparing dinner; I was folding washed clothes. Queen's sound waves throbbed through the room. Freddie Mercury captured the essence of his roots from Africa and India, his transition to the UK and his challenges with racism and homophobia in his vibrant music. We had watched 'Bohemian Rhapsody' on the big screen in Bordeaux the week before. Freddie touched our hearts with universal love, something that connected all things.

I put my hand on the antique pine table, vibrating like a living thing, transmitting sound from the mobile speaker.

Sophia had started boarding school in Bordeaux. She was independent, organising her life, taking the train into the city on her own and navigating the place like she was born there. She came home on the weekends. That Friday she returned home in an antsy mood, and I ramped up my antsy to meet hers.

Sean took me aside and reminded me to be a mum. I sometimes forgot that. I forgot what it was like to be a teenager. I was still struggling with my own dreams, desires, and demons. Sophia and Ellie needed me to be an adult, to support them.

Freddie said this place could be heaven for everyone. He was right, it could be, and at times it was. Sophia calmed. I settled into thankfulness, for food in my belly, family around me. Dora, our dog, twirled in her bed looking for the perfect spot. She was eleven, sixty-five in dog years, and still twirling. I took her out for an evening walk. Jupiter, Orion's Belt, and the Plough were bright in the sky. The poplar trees were playing music in the breeze, a gentle flutter, like paper and fairy music.

When I returned, Ellie was making Freddie's sounds that connected us to the universe on the piano. Sean was chopping kale. Sophia was peeling potatoes. I felt magic in the cosmos. Like Lucille Clifton in her poem 'Cutting Greens', there are musicians, artists and poets that can take us beyond, into the wonderous mystery of life in simple daily tasks. Finding awe in the everyday and feeling the welling of bliss from these things. That is true art.

CHAPTER 13

When is a prairie not a prairie?

I caught sight of Stonehenge on my way to Sherborne. The circle of stones on Salisbury Plain in Wiltshire, England, was constructed more than 4000 years ago. It is oriented to align with the sunrise on the summer solstice and has a timelessness, a spiritual awe. I felt a connection to the past through it. Perhaps I had been reading too much Diana Gabaldon. Diana wrote the historical fantasy romance Outlander book series about Jamie Fraser and Claire Randall, wherein some of the characters time travel via magic in standing stones similar to Stonehenge.

In Sherborne, in keeping with the history theme, I slept in a four-poster bed from the 1700s and attended a music recital in an abbey dating to 1300. I was there to speak at the Sherborne Festival, a kind invitation from the Sherborne Abbey Festival, to share our story and organic message, to sell wine and books, and be cocooned in a luxury hotel.

But it wasn't the historic buildings, the grand Abbey, the historic town centre, or the event that left their biggest mark on me. It was trees. On a run to the hill behind Sherborne Castle, a clutch of ancient oaks felt like a fairy glen. I walked and danced, me and the trees, delighting in each other, in being together. They sensed me, and I, them, at a deeper level than sight, our atoms vibrating with each other.

Keen to see what was around the corner, I ran onwards, through a field and into a young oak forest. I didn't feel them like I had the old trees. The wind rose, bringing a bank of dark clouds scudding towards me. I turned back. As I passed the old oaks on the descent, I lifted my hand and caressed their low hanging leaves. I felt a love flow between us. I wanted to stay, but the cloud was menacing. I gave them a gentle farewell caress, sad at leaving, then ran down the hill. Rain pummelled me by the time I reached town. I drew a hot bath and thought of the massive oaks as I soaked, of the ancient wisdom they represented and how important they were. Soon after I would read 'The Hidden Life of Trees' by Peter Wohlleben and realise that there was even more wisdom in trees than I thought as I daydreamed in that tub.

From Sherborne, I travelled to friends near Oxford. Mike shared a recently published report on biodiversity[12], by the Intergovernmental Science-Policy Platform on Biodiversity and Ecosystem Services (IPBES), an independent intergovernmental body initiated by the United Nations. The report estimated that one million species including a quarter of all mammals, were at risk of extinction. The threat that accelerating loss of biodiversity posed, and the challenge it represented for us humans, was, according to the authors, on a par with the climate crisis, and directly linked to it. The three main drivers of decreasing biodiversity were habitat loss, climate change, and pollution.

Our small blue planet is a massively complex biosphere. All species are interconnected and often depend on each other. One loss may not seem important, but it has a knock-on effect, and the losses were accelerating. The air we breathe depends on plants and much of the food we eat depends on pollination. We need them more than they need us. The report outlined that we were in the throes of a modern, man-made, mass extinction, the likes of which had not been seen since the demise of the dinosaurs, when three-quarters of the plants and animals on earth were wiped out by an asteroid crashing into our home.

Elizabeth Kolbert outlines these facts and more in her Pulitzer Prize winning book, 'The Sixth Extinction, An Unnatural History'. Industrial farming by monoculture was not helping. Depending on single species intensively and chemically farmed, as humans do with many of our foods in the modern world, is not a robust way forward. Monoculture means less resilience in our food system.

We think of the usual suspects of climate crisis and pollution, but many of Kolbert's examples were even simpler, through human transport and travel. Our goods, and ourselves, transfer pathogens from one continent to another. Like the story of *Flavescence Dorée*, where humans transferred the virus vector American grapevine leafhopper from the Americas to Europe, she outlined how the recent mass die-off of frogs could be traced back to a fungus. The fungus was present on African clawed frogs shipped all over the world in the 1940s and 1950s for use by obstetricians in human pregnancy tests. When chemical tests were developed many doctors set the frogs free not realising that they were creating a frog Armageddon. While the African clawed frog could live with the fungus, many other continents' frog populations were not able to resist it.

I had recently read Kolbert's book and another on biodiversity, 'Wilding' by Isabella Tree. Tree's book was one of hope, about how quickly biodiversity took off after she and her husband Charlie Burrell transformed their intensively and chemically farmed 3500-acre property in Sussex to wilderness farming, with hardy breeds living as they would in the wild. Their rewilding story was fascinating and uplifting.

On our 35-acre farm in France, we had also seen hope-filled results for biodiversity. After three years of organic farming, many species of wild orchids including two types of pyramid orchid, the tongue orchid and two types of bee orchid, returned. This was because the systemic fungicides had worked their way out of the soil and the mycorrhizae had returned. Orchids are dependent on mycorrhizae, a special fungus that develops on the root systems of 95 percent of plants tested to date and helps them access nutrients in the soil.

This mutually beneficial relationship came about more than 400 million years ago. Research shows that mycorrhizal networks transport carbon, phosphorus, nitrogen, water, and defence compounds – compounds that combat attack by something, from plant to plant. They are especially beneficial for plants in nutrient poor soils. The best wine grapes are grown on relatively poor soil, so it is particularly interesting for us winegrowers. To understand more about this relationship, I recommend the fascinating autobiography of Dr. Suzanne Simard, one of the first scientists to research this phenomenon, 'Finding the Mother Tree'.

These magic mushrooms will not develop in chemically farmed soil because of systemic fungicides. The wild orchids that appeared after three years of organic farming at Château Feely, were an indicator that the farm was clear of the systemic fungicides that had been used before we arrived in 2005. Systemic fungicides are bad for us, they can be carcinogenic (cancer-causing) and nervous system disruptors. They are 'systemic', which means they go inside the plant and the fruit and can't be washed off. Biodiversity of plants, insects, and small mammals grows when we farm naturally.

We have set aside more than twenty percent of our farm as a wilderness area. We also allow wild plants to grow between our vine rows creating plant biodiversity that brings insect biodiversity. Through insect biodiversity, we have a large population of good bugs like ladybirds, Typhlodromus, and common green lacewings. These beneficial insects keep the bugs we don't want, like aphids and spider mites, at bay.

That year, to build awareness of organic and biodynamic farming, Ros, our apprentice studying wine tourism development, and I, created an organic and biodynamic trail for the farm, a physical trail in the vineyard, and a virtual trail on our website. It was a way to share our experience and cultivate change beyond what we were doing on-site. The scale and challenge of the climate crisis and biodiversity loss are immense. We all have a part to play – and we must *play* – this can be a fun transition, one that celebrates life, not one of fear. As humans, we need to feel like we are moving forward, that we have a common objective, and that we are part of the wondrous interbeing. Every step we take from rewilding our gardens, encouraging abundance on our farms not weedkiller, choosing organic products, saying no to plastic, reducing our carbon footprint or joining an environmental movement, is helping to turn the tables.

Many individuals are realising the scale of these challenges, but our institutions did not seem to be. The EU and its agricultural policy weren't keeping up with biodiversity or climate change. I watched a brief video about a farmer who had had his Common Agricultural Policy (CAP) aid payment withheld because he had trees in the pasture fields where his cows grazed. The CAP required it to be grass only. With global warming, shade for animals was more important than ever, and trees provided it. He filmed his exchanges with the bureaucrats and created a video that was worthy of a John Cleese spoof. I laughed but was also scared by it. I congratulated myself that we hadn't had any problems with the small, but appreciated, organic support payments we received through the CAP.

The universe laughed and a few days later I received a letter saying that due to anomalies in our CAP submission three years before, the amount we were to be paid was zero. Lead plumbed my stomach. We were counting on receiving the outstanding sum, a backlog of more than three years of payments. We had a loan for winery equipment that depended on it. I ran across the courtyard to find Sean.

'You'll have to phone them,' he said. 'I'm sure it's a misunderstanding. We're certified organic and they need to pay us the maintenance aid.'

We had met bureaucratic barriers before, and they were sometimes illogical and inflexible. Sean filled in our CAP forms, but since his French wasn't good enough to make phone calls, I had to be his go-between. It was 6.30 a.m., and their phone lines didn't open until 9 a.m.

I found it hard to concentrate and eventually gave up working on the PC and did some weeding outside the tasting room instead. At nine, there was an hour until guests arrived, hopefully long enough for what was usually an interminable wait for a helpline. Miraculously I got through almost instantly.

Madame was patient but adamant, the anomalies in our application meant no payments. I wasn't sure what Sean had done and felt out of my depth. I asked her to hold and ran across the gravel courtyard again, my heart pounding. Sean was at the kitchen table filling in his vineyard file. With my hand over the mouthpiece, I quickly explained what she had said.

'Ask her what the anomalies are,' said Sean.

I translated his question and then the answer.

'She says you declared the open fields as prairies, and prairies have to have animals on them,' I said.

'But we didn't declare any animals, so of course they don't have animals on them. I thought a prairie was a field of grass.'

'She knows, but under CAP, what you declare as prairie has to have animals on it. It's the definition of prairie for CAP.'

Sean looked flustered. The voice on the other side offered another stream of fast-flowing French.

'She says we have done the same thing for three years, so we won't get any aid on the other years' applications either.'

'If they had processed our submission when they should have, we would have been able to fix the error quickly and simply, and we wouldn't have made the same mistake on the following years,' said Sean.

I translated what he had said. She became a brick wall. My heart raced, and I felt anger flare.

'You may even have to pay a fine,' she said.

Fear overtook the anger. Not getting the aid, due to a tiny misunderstanding was bad enough. We had to play ball and find an amicable solution. Madame offered to send us a detailed email explaining the anomalies.

Prairie had to have *'cheptel'*, grazing animals, on it. If it was prairie without animals, it had another name, *'friche'*, or something else. There were many potential terms, even Madame didn't seem sure which we should opt for. In addition, she pointed out that under the CAP organic maintenance agreement, we were not allowed to change anything on the farm for the five years of the programme. The old, unproductive part of the vineyard we had grubbed up a couple of years before, was also an anomaly. It was illogical to ask farmers to keep fields static for five years, but it was how CAP required us to be.

The greatest irony was that we had booked help from the Department of Agriculture to do our CAP submission after the grubbing up, knowing that it changed the picture. It was thus one of their personnel who had helped us set up this erroneous submission. After a few emails, Madame suggested we come and see her.

I called our friend Isabelle, wife of Thierry, who had helped us with his *Cirque du Soleil* winery moves. She knew the CAP. She was a teacher at an agricultural college, and she did her own farm's submission each year. Isabelle was almost as confused as we were about the *friche* versus *prairie* and the potential ways to describe resting fields and wild areas. CAP was complicated.

A couple of days later we drove to the *cité administrative* in Perigueux, the bureaucratic heart of our department. After waiting briefly at reception, we were welcomed by our contact. Madame was fit and tanned with short hair and glasses. She gave off a friendly, can-do attitude. As we started the meeting, I felt confident that we would work something out.

Fifteen minutes later things had gone nowhere, and I wondered if we would be leaving with no aid and a fine. Then the fire alarm went off. We followed Madame through a warren of corridors and down the steps to the grass outside. It was late May with an unusual polar wind and the smell of rain. Outside, in the flood of evacuated office workers, we felt exposed in every way; cold, out of place, and at her mercy.

After a long wait and a roll call of the representatives of each floor, we were allowed back in. We sat back down in front of her desk, like errant schoolchildren. A massive flat computer screen took up a third of her desk and the other two-thirds were filled with piles of *dossiers*, folders of PAC applications for farmers in the region. She called our farm up on the screen, zoomed into the detailed aerial photos, and interrogated us on each part and what it was.

While everything was in the spirit of the programme, we were maintaining our organic farm, and our errors were simply misunderstandings, at that point I felt all was lost. I would have been happy to walk away empty handed, so long as she didn't levy a fine on us. We would find a way to pay the loan that was depending on the aid. It felt unfair. But I could also understand that as an administrator of the CAP she had to follow the rules laid down by the EU. I recalled reading that Spain hadn't followed the letter of the law on CAP and had paid heavy EU fines for it.

In those moments of acceptance of what would be, there was a stillness in the air, the rain stopped, and the polar wind abated. Madame looked over her screen, a great digital rendition of our home, a tiny piece of France that we had been tending back to health for more than a decade.

Sean and I waited for her ruling like sparrows watching a hawk. She stopped looking at the screen and faced us. She began to debate options, ways we could solve the issues without getting her, or us, into trouble. I felt déjà vu, a sensation as we had in the Bergerac customs office soon after we arrived in France. That time, a crowd of people got involved, the massive customs code book was consulted several times, and they debated what to do with us looking on for about two hours. It felt like a long time since breakfast, but I would have to hold my hangry. Madame continued debating with herself for a while. At the point where I expected her to call a crowd of colleagues, she offered us a handy way to fix the situation and proceeded to help us to do it.

I have no idea what tipped the balance, but soon after I hit a position of acceptance, of letting the meeting go where it wanted to go, there was a shift. We left feeling upbeat, sure that we wouldn't have an additional fine, but unsure what the level of aid, if any, would be. A couple of weeks later support payment arrived in our account in time for the loan repayment.

As we found in our early days in France when we had regular scuffles of misunderstanding with bureaucracy, people were generally helpful, professional, and on our side. We thanked the sky and sent thoughts of gratitude her way. Feeling high on the news of the aid payment, we booked a holiday to Italy; something we had been dreaming of for years. I sent Steph the dates in case she could make it. It was the year of her 50th and Russ would have been fifty too.

One of the anomalies in our CAP submission was the area we grubbed up, where we planned to plant the new Chardonnay. It was time to finalise the vine order and pay the balance. But, given the speed global warming was advancing, we were no longer sure that Chardonnay was a good choice. Sean was also keen to have disease-resistant varieties.

He had tested hybrid vines that were resistant to downy mildew in the garden. They were resistant. But we bought and tasted wines made from locally grown hybrids, including the ones we were testing, and didn't like them. Since placing the Chardonnay order, our thinking had moved on. Looking at projections for global warming, our farm would soon be too hot for great Chardonnay.

Given how long vines live, we needed to look out about 80 years, so it was a tough call. There was much more to assessing the conditions than a general temperature increase. Frost was a major consideration. We couldn't consider vines that were precocious as they would be more susceptible to frost. Winegrowing was a long and complex game.

We read research papers on the subject. One thing became clear. Chardonnay was not a good choice. High-quality Chardonnay was suited to a slightly cooler climate than our current grape varietals Sauvignon Blanc and Semillon. All the predictions were towards hotter conditions. We had been wooed by the taste of a magnificent local Chardonnay instead of considering what they would be like in a hotter future.

The following day I cancelled the order. As if to confirm our decision, summer rolled in with several peaks of over forty degrees Celsius. Sean was out clearing weeds at 10.30 a.m., and it was already so hot that crossing the courtyard from our kitchen to the tasting room was a challenge. As I looked out of the window, I thought of going to tell him to get inside. The memory of my heat disintegration on the Monbazillac walking tour still brought a frisson of angst. Then I said to myself, 'he knows his limits'. I also didn't want to face walking to where he was working, crossing the fifty metres felt like mission impossible. That should have been the cue for me to force him inside, but I didn't pick it up.

Ten minutes later he came in. He was red and feverish and had a migraine, nausea, and cramps. It was heatstroke, a condition that can be fatal if not addressed swiftly. Sean followed Doctor Internet's instructions. He took a cool bath, drank plenty of water, and lay in bed. His headache was so bad he couldn't read. He lay in a dark room with the shutters closed and the fan on, drinking water and taking paracetamol. He was seriously sick for two days.

Inside the house was a 'cool' 25 degrees Celsius with no air conditioning. We opened everything up at dawn and then closed shutters, curtains, windows, doors, before 9 a.m. By keeping them closed, and keeping the sun out, we kept the house cool naturally. A couple of days later it was even hotter. Clients staying in the Wine Lodge told us they had closed their factory in Oxford, England. It reached 38 degrees, a record in Oxford, and their server had shut down. It was too hot for machines and for humans. Everyone went home.

In Saussignac, by 11 a.m. there was no sound outside. No insects or birds were moving. Despite being in a shaded area and having water and shelter, one of our chickens died. I felt even more motivated to spread the word about the climate crisis and the need to reduce carbon dioxide emissions. I signed up to the organisation 350.org and to Greenpeace France. The climate crisis was serious and real, and it was coming at us faster than anyone expected.

Sean recovered and a minor family dilemma took my mind off the climate crisis. Sophia had arrived home Friday evening and we chatted as we cleaned up.

'I want to get a tattoo,' she said.

'I don't think that's a good idea,' I replied. 'You want to be a doctor. Do you think people will trust a doctor with tattoos?'

'Of course, they would,' said Sophia.

'You might think it's a good idea now, but imagine later in your life. Your ideas will change, and the tattoo will be permanent.'

Ellie watched the debate, interested to see where it would go. She didn't put much store in what we said at the best of times. Sean came in and I updated him on the conversation.

'It's your decision,' said Sean. 'But we're not paying for it.'

I went out to the office to lock up. When I came back, Sophia had found research that said it was okay for doctors to have tattoos. Good for her for proving my prejudice wrong. I chuckled as I went upstairs to pack my bags for Italy. What would be, would be. If she went ahead and regretted it later, it was her mistake to make, not mine.

CHAPTER 14

Awakening in Cinque Terre

Our taxi driver had no English and no brakes. I gripped the edge of the seat and prayed as we flew along a five-lane highway in Milan, low-slung apartment blocks whizzing by. Our temporary home was a basement apartment in the Lambrate neighbourhood, owned by our beloved house-sitters, Crowded Planet travel bloggers Nick and Margherita's family. It was packed with thought-provoking books; home to a publishing house run by Margherita's sister.

Milan had big city energy. Graffiti covered almost every street-level surface except the *Duomo*. The Milan Cathedral, or *Duomo* in Italian, is more stunning in reality than in photos; the pale pink marble reflects a luminosity impossible to capture. Near the entrance, we found a grape harvest sculpture, then a dragon wreathed in a forest.

We were pushed forward for a body search with a metal detector. Cleared for access, we followed winding stairs to the rooftop where a crowd of soaring statues took our breath away.

Milan is history and art, but also fashion. Galleria Vittorio Emanuele II is a famous arcaded mall adjacent to the Duomo. It is packed with luxury brand shops and a magnet for style bloggers. They thronged through, taking selfies and spinning on the bull's balls in the mosaic at the centre of the mall. Spinning anticlockwise on your right heel three times is supposed to bring good luck. Thanks to daily torture by hundreds of tourists, all that was left of his balls was a well-worn dent. I was one of them, the only person in our entourage to succumb.

The Galleria hosted upmarket Italian brands, Prada, Gucci, Massimo Dutti; but on the affordable high street, it was international chains, the same as could be found in Bordeaux, London, Dublin; usual suspects like H&M, Zara, Pimkie, Footlocker. We could have been anywhere.

We rode old trams, wooden benches reminiscent of an ancient yacht, and art deco lights. It felt like a film set for the 1930s. From the smooth varnished seats, we stepped onto the pavement and found 'God Save the Food' for the best fresh juice and coffee we'd ever tasted. The anti-inflammatory health shot, a burst of lime, carrot, apple juice, ginger, and curcuma, was a blast of well-being. Sophia had researched and found all the places she wanted to visit. Sean wanted to do museums. The Milan Museum of Science and Technology, also known as the 'Leonardo da Vinci', was high on both their lists, so we started there.

The world queues to see da Vinci's Mona Lisa in Paris, but the room filled with his inventions reminded us he was more than a great painter. Da Vinci was a master of biomimicry. As a child, out on the river in a small rowing boat with his uncle, they noticed a frog and watched him swim for a while. His Uncle said, 'If you do what he is doing you can swim'. Leonardo pointed to a bird flying overhead and said, 'If I do what he is doing I can fly'.

His uncle said, 'Maybe not,' or something like that, but it stuck with Leonardo and informed his flying machine creations later in life. He observed how birds flew; how a seed or leaf floated in the air. A tree's seed that flew like a helicopter, perhaps a sycamore seed, was the inspiration for his helicopter plans.

Many of his inventions were only progressed 300 to 400 years later, like the steam engine and the helicopter. The museum was massive, and long before we had finished exploring, we were out of time and energy. With planning and food, you could spend days in that museum and not reach the end of it. We needed gelati, a scoop of dark chocolate sorbet and another of fresh lemon. No wonder Italian ice cream is world famous.

Sophia picked the Brera Pinacoteca art museum as the main feature for the next day. I was sceptical, not keen on a boring art museum. I was so wrong. The art was breathtaking, but more than that, each major work had an explanation of why it was so important in the world of art.

'Melancholy' by Francesco Hayez, painted in oil on canvas around 1841, showed a girl and a vase of fading flowers, her sleeve slipped off the shoulder. It was one of the first times that mood was shown so clearly in a painting. I felt her melancholy like it was mine.

The notes asked us to reflect on things like: how would you seat someone sitting for you for a painting? What makes a masterpiece iconic? Often, it was a new way of showing a subject or approaching a story. The Marriage of the Virgin by Raphael, oil on panel, created in 1504, showed perspective to perfection for the first time. I felt like I could walk into it. It also took a Bible story and recreated it in clothes from the era of the painting rather than that of the protagonists.

Perhaps the most remarkable was 'Jesus Last Supper' by Rubens. Judas looked out of the painting like he was alive. I felt his guilt like it was in the room with me, despite that painting was completed in 1632 and depicted a story from nearly two thousand years before. The emotion poured out of it.

The Brera area and the art museum alone made our trip to Milan worthwhile. Being with locals Nick and Margherita in the evenings made our experience even more unique. They took us to 'da Zero', rated number 15 in the top 25 pizzerias in Italy, and the only one in Milan on the list. The restaurant was proud to share photos of their suppliers for flour, cheese, tomatoes and olive oil. Everything was homemade and locally sourced. The truth was in the taste.

Months before, when we debated where to go for the other four days of our holiday, I put Cinque Terre, on the northwest coast of Italy, as my top choice. The photos shared by hipsters on Instagram convinced Sophia and Ellie it was the place to be. A high-speed train whooshed us out of Milan into the flat fields of corn and farmland of the Po Valley, and then into the mountains, before dropping down to Genoa to follow the coast. I hoped Cinque Terre would live up to expectations. Sean and I were making good progress with our relationship rebuild, but he wasn't convinced about the choice of Cinque Terre. I didn't want another situation like in Dublin, where he didn't like my choice of apartment or location.

Skittering views of Genoa, turquoise sea and colourful Ligurian apartment blocks and mansions, made me want to stop, but our destination was Monterosso, the mainline stop on the train track that ran through Cinque Terre. From there we would train on to Vernazza where our apartment was. It was too early to check in, so we decided to lunch in Monterosso. We descended from the station in a scrum of holidaymakers.

Online articles had said to beware of summer in Cinque Terre because of the heat and crowds. I could see what they meant. Sweat poured down my face and the suitcases ripped my shoulders out of their sockets. Overloaded with bags, overwhelmed by heat, crowds, and two unhappy adolescents, I wondered if this destination *was* a mistake. Sophia and I were prone to attacks of hangry and we both had one coming on. I started to feel that same dizziness I had on the roadside of Monbazillac. We were saved by a charming restaurant with a shady terrace clinging to the hillside between the station and the main town.

The view over the sparkling sea, friendly staff, and a chickpea *farinata* with Genovese pesto, mozzarella tomato salad, and gnocchi, stopped the hangry and transformed our vibe. Sated and revived we returned to the station. As we waited for our train to Vernazza, a 5-minute hop away, I heard a group of tourists recently landed in La Spezia on a cruise ship grilling the ticket agent about options for the two hours they had in Cinque Terre. After a quick review, they bought tickets that would allow them to stop in each village, get out, walk to the main street, and then get back on the train. It was fast tourism, like fast food – filling, but not satisfying.

Vernazza, our destination, was the second town of Cinque Terre heading south, a layered port village surrounded by sea cliffs. We poured from the train in a posse of day trippers and flowed to our contact at a travel agency on the main street. The multicoloured buildings and pedestrianised streets were more beautiful than I imagined, but I felt a nip of worry about crowds and heat.

Our hosts welcomed us, happy to check us in early. By the top of the six steep flights of stairs to our apartment, my thighs were screaming. We'd get fit going in and out each day. The Wi-Fi didn't work and the small, typed note next to the internet box said; 'the service here is not good, there is nothing we can do if the Wi-Fi isn't working, don't bother to call us.'

I put the kettle on, to make some tea, and the electricity tripped. I walked downstairs wondering what to do. In the entrance landing an ancient electricity box was niched behind the door. I pulled the cover and saw one switch down. I pushed it up and climbed back up the six flights. Eureka.

We realised that we should not overload the electricity. It could support one major appliance at a time. If the washing machine was on, the kettle could not be. Very ecological. The high ceilings, wood floors, large balcony looking onto the hills, and, as we would discover, a small beach less than fifty metres away, more than made up for it.

The girls collapsed on their beds. Sean and I went to suss out the village. The entrance to our apartment block opened onto the main street. It would have been an issue in any other seaside town, but this coastal gem had no car traffic, part of the secret of Cinque Terre. The villages were only accessible by boat or on foot until 1870 when a train line between Genoa and Rome was built that passed through the towns. The difficult terrain and sea cliffs meant that roads did not reach them until the 1960s. Even now, cars are not allowed in the villages. Each village has a parking lot near the top of the village, accessible from the coast road, but not recommended.

Sean and I followed the start of the hike we planned for the following day, to get a sense of the lie of the land. As we climbed, elegant houses and cobbled streets were replaced by terraced vines. Before turning back, we sat for a few minutes to appreciate the view of the multicolour village and shimmering sea. We felt at home. Something in it reminded us of the Western Cape province of South Africa. We held hands and breathed in molecules redolent of wild herbs and iodine.

Back in town, the day-trippers had departed, and the streets were quiet. I rallied my family for a swim. A tunnel adjacent to our apartment entrance led to a small beach. Sophia, Ellie, and I swam in the shallows; Sean took off into deeper water. I wanted to go but I wasn't brave enough.

Swimming with Russ near Durban in my teens, we were pulled out by the tide. I felt like there was no way I would get back to shore, my strokes no match for the powerful backwash. Russ powered back and ran to call the lifeguards. I got the better of the tide before they arrived. The experience left me exhausted and with a new level of respect for the ocean. At the Eastern Cape beach that we visited with my parents, Steph had a similar experience.

Floating on my back, I thought of them and wished they were with me. I felt a wave of missing Russ. He reminded me to grab life with both hands, as he always had.

I rolled onto my front and swam into the gentle waves feeling the pure bliss of the sea, then rolled onto my back, relishing their rise and fall, the gentle sensation of them pushing me into shore. When I turned back over, I noticed Sophia and Ellie had swum out. I called them in. As I did it, I realised I should not have. I had to let go. They needed to experience things for themselves. I couldn't stop them from taking risks and discovering life. Sean was out there if they got into trouble. They needed freedom. They would soon be adults out on their own.

That evening at '5 Terra Bistro', a short walk up the main street, a golden orange wine from Vétua in Monterosso was rich with candied peel, fruity and dry, perfect with seafood. I read they were a small, family-operated, naturally farmed vineyard that produced 4000 bottles a year, an artisanal winery even smaller than Château Feely. After dinner we walked down to the harbour, a saucer moon shining brightly on the calm sea.

I was reminded of the poem 'Drinking Alone Beneath the Moon' by a Chinese poet of the eighth century, who is commonly referred to as Li Bai or Li Po. Alone, he pours a glass, then makes a threesome with the moon and his own shadow. They sing and dance, then scatter to reunite sometime in the future with the stars on the far side of the Milky Way. Merely thinking of it made me feel like singing and dancing, but instead, we bought *gelati* and sat on the harbour wall to enjoy the sounds of lapping waves, people's voices, and Italian ice cream.

Raising teenagers from their beds before midday was tough; getting them up at 6 a.m. on holidays was madness. By seven, crazy as we were, we were on the trail to Corniglia, lamentations ringing, as we toiled up the steep start.

'I don't want to do this,' said Ellie.

The trail flattened. Vistas of hilltop villages and azure sea opened around us. The lamentations were replaced by silent wonder interspersed with exclamations of delight. We stopped to sit on a rock, in stillness and silence, to take in the joy of it.

Further on, olive groves of matt grey-green smattered between bright vineyards and scrubby wilderness. I wanted to do the whole coastal trail that connected the five villages, but a mudslide a few years before had closed many routes, and two were still closed. Mudslides were becoming more frequent with global warming, the result of extreme rain, sometimes made worse by the conversion of land from native vegetation or forest to farming. We passed a handful of walkers then a bronzed local, coming up to his home in the hamlet of Prevo fresh from a swim in a secret cove below.

Our destination, Corniglia, had around 200 inhabitants and an even slower pace than Vernazza. There were fewer day trippers than in the other villages due to the steep path up from the station. We loved walking into it from the north, following a small, paved route, too small for cars. An agitated voice in the vineyard above us rang out. Davidé, his phone correspondent, hadn't turned up for work and perhaps wasn't going to. I couldn't blame him; the terraced vines were so steep that tractors could not work them. Mechanical pulley systems served to bring equipment up and harvest down.

We found a friendly café and ordered coffee and snacks. It was the best coffee we had ever had, even better than 'God Save the Food' in Milan.

'I want to do this again tomorrow,' said Ellie.

Boosted with caffeine, we trained back to Vernazza to swim in 'our' alcove beach. I bought focaccia and fresh tomatoes and we retreated to our apartment. Crowds from the cruise ships in La Spezia and Genoa arrived around 11 a.m. and left by 5 p.m. That was the hottest part of the day and the ideal time for us to hunker down for a light lunch and siesta at home.

The next day I queued at a stand selling cheese, pasta, and fresh vegetables, well attended by wizened locals. Each time a local joined the queue, the artisan would serve them, and leave the rest of us to wait an extra turn. I felt put out, then realised he was doing the right thing for them, and himself. These people supported him all year round, us tourists were there for a brief stay. His mozzarella was worth the wait.

Later, our hike took us to the Sanctuary, 45 minutes of steep ascent out of town. It was where the villagers would go when they were attacked from the sea in medieval times. We needed the shaded Sanctuary when we reached it in the heat of late morning. After drinking litres of water and recovering in the shade, we scampered down, sweat balls desperate for a swim, but our beach paradise was packed with tourists, both in and out of the water, and the waves were rough. We dove in to cool off but didn't linger.

Late afternoon the beach was still crowded, and the sea was wild. Sophia and Ellie decided they didn't want

to swim. We gave them the keys and took to the sea in
the harbour instead. Dripping and ready to shower in
preparation for our dinner reservation, we pushed on the
apartment block entrance door and found it locked. It
was always open. Sophia and Ellie were three stories up
and unlikely to step onto the balcony. We didn't have
phones, there was no intercom, and we somehow knew
they wouldn't notice the time, or even consider we were
locked out if they did.

We sat on a step on the opposite side of the street, where
we had a view of the entrance and our balcony in case our
daughters appeared. We didn't know their phone numbers
to call them if we could find a friendly passer-by, a danger
with modern phones where their memory has replaced
ours. We sat waiting, wondering when they would realise.
The restaurant reservation was the only one I had made
months before. I didn't want to miss it. A half-hour later,
as I considered walking to the restaurant to see if I could
push our booking out, another guest opened the building
door. We shot up and begged to be let in. In minutes we
were speed showering and laughing.

We felt like rock stars with our table above the crashing
sea and views to forever. The seafood was fresh from the
local boats, unlike some of the restaurants on the main
square. The restaurants in Vernazza noted whether the fish
was fresh or frozen, and if it contained sulfites, so guests
knew what they were getting.

As we wandered back, music drifted down to us from up
the street. We followed it. A little beyond our apartment, a
band had set up an impromptu concert on a small square.
They were talented, beautiful, and full of energy. I found
my feet tapping and my body swaying. Sophia and Ellie

couldn't believe that they had stumbled onto a free rock concert by a group of cool music students in some hiking destination chosen by their unhip Mum. We clapped and danced until the village decreed silence in the land. A hat was passed around for donations, and we wandered home in a glorious afterglow.

In the morning, Sean and I left our daughters to sleep in and took the Azure trail from Vernazza to Monterosso. The path started northwest along the head above the harbour. Prickly pears, their orange-coloured fruit, like fat fingers, attached to large oval succulent leaves, framed the sea. Waves crashed on the cliffs, iodine aroma mingling with wild herbs. Around the corner, magenta bougainvillea tumbled along a fence protecting a small vineyard that clung to a slope between the path and the cliffs.

We stopped intermittently to gaze at the wild seascapes banded by rustic olive groves and local scrub. As we got closer to our destination, olive groves were replaced by terraced vines, including those of Vétua, the wine we enjoyed on our first night. The steep vineyards looked well-tended and loved. Beyond them, the trail descended into the outskirts of the town, and to a small stretch of beach open to the public but empty of people. Most of the beach in Monterosso was owned by hotels or clubs, and guests had to pay to be there. We changed into our swimming gear and luxuriated in the cool ocean. By the time we were towelling down at 9 a.m., ten people had staked out spots on the sand.

Our days melded into early morning hikes and swims, afternoons at home, late afternoon swims, and dinners out in vibrant evenings. I discovered locally made fig and almond gelato and decided Vernazza was my spirit home. Sean and I rose early, the promise of a swim pulling us together like a tide. We sipped coffee on the balcony and soaked in the mountain view. It reminded us of the mountains in the Little Karoo in South Africa. We felt at home, like the geography here was a confluence of the Europe we had come to love, and our original homeland. Towels flapped in the breeze; the sounds of people's voices floated up. Every apartment had laundry lines hanging out on the terrace and in front of windows over the alleyways out back, offering colour and life.

Hand in hand we walked down to 'our' beach. It was empty, truly 'ours'. The sea was calm. With Sean near me, I felt confident to swim out, beyond where my feet touched the sand. Out on the deep blue rollers, I looked back and drank in the view of the cliffs and the back of the village. My heart sparked with happiness as I floated on my back, the waves lifting me up and then drifting me down. Sean's hand touched mine and I took it gently. We kissed, then floated away, fingers touching. I felt so ecstatic, I could have exploded right there, and gone to heaven.

That moment I knew our relationship was healed. We had made it to the next phase. We loved each other, not the fizzy love I felt the weekend Sean proposed, but a deep love and understanding of life partners, a love that included but went beyond sex. It was a spiritual bond, a knowing of each other like no one else.

CHAPTER 15

The earliest harvest ever

T hanks to our amazing apprentice Ros, we had taken our holiday to Italy without worrying about the business. She was reliable, knowledgeable, and loved by our clients. Ros had passed her wine tourism diploma and was planning to go travelling to New Zealand like our previous apprentice Cécile had done. We would be sad to see her go. It would be tough to replace her. I put feelers out to the colleges that offered tourism and wine tourism diplomas around us.

Apprentices were a good solution for us given how seasonal our visitors were. An apprentice was with us for about half their time; at school for about a third of their time and on holidays for the rest. We paid them a fixed percentage of the minimum wage depending on their age; but did not have to pay the usual level of social charges, a major saving in France. I couldn't cope without a helper, particularly in the season, from June to September. I waded through CVs but none of them had a good enough level of English. We needed someone fluent in French and English, and keen on organic farming.

Given we were located over an hour from the tourism schools, we were not ideally placed to attract candidates. For most of them, an estate in Bordeaux was more attractive, despite our tourism awards. I was close to giving up hope when an email arrived from Elodie. She had attended school in Holland and France and had worked in Ireland for two years. Her Dutch French parents ran a hotel, restaurant, and camping business, where she had worked many summers, and she had recently worked several months at a Michelin-starred restaurant. I could not have dreamed up a more perfect candidate.

With Elodie set to join us, my side of the business was covered, but Sean needed help with the winery work. His back was still tender. I couldn't face seeing him in the devastating sciatica pain again. If it happened earlier in the harvest season, we would be in trouble. In a miracle of timing, our nephew Duncan wrote to say he had a month free before starting work and was thinking of coming to help with the harvest.

In the first few years in Saussignac, harvest started mid-September and finished mid-October. Now we started in August. With hotter summers and global warming, grape picking crept earlier and earlier. Sophia and Ellie were still on holiday and helped pick the whites and rosé. Having our daughters bolster the team was great, but an early harvest meant managing peak season guests in the accommodation and tasting room, and harvest, at the same time. Ros and I raced from vineyard to tasting room quickly changing our looks from grape-stained and sweat-smeared to smart and professional.

We had a GP in place at last. I didn't want another broken back with no doctor. I took Sophia to sign up with him as her *médecin traitant* before going back to school, singing his praises en route, how thorough he was, and his great manner. He was professional, humorous, and compassionate. After a short spell in the hot waiting area, we were shown to his consulting room. Short sleeves revealed a large tattoo over most of his forearm that hadn't been visible when I first met him in winter. Sophia muffled laughter as I smiled guiltily. On the way home, we laughed heartily at how well my prejudice had been shot through. *Mea culpa.*

Elodie and Duncan took the baton from Sophia and Ellie when they went back to school. Elodie would have a month with Ros to learn the ropes as we faced our busiest September ever with large groups, back-to-back tour days, and harvest. Ros and Elodie were instant friends, working naturally together. I loved the buzz with them and Duncan. We moved through the dance of picking grapes and winemaking like a perfectly choreographed ballet. We did harvest, tours, and maintenance, and stretched out with creative photo shoots. As well as her other skills, Elodie was a keen photographer.

To complete the harvest circle, Steph came to visit for a week. We were getting our 50th celebration at last. We picked the last tiny section of Merlot grapes.

'It's so good to have you here,' I said, tears in my eyes.

'It's so good to be here. Thank you, my wonderful friend.'

We toasted Russ with wine and memories, walked the paths of Saussignac and Monbazillac, catching up and reminiscing, then Bordeaux city for the weekend. When I got home after saying farewell to my longest-standing friend, the lone oak down in the hedgerow below the house waved in the wind like he was talking to me. He was lonely and more at risk than grouped oaks. In big winds, he had to resist alone. He didn't have others in his tribe helping him via roots connecting and the wood wide web of mycorrhizae. Like us, he needed friends and family to be strong.

My voice went husky as I wrapped up the last day tour of the season. By the time I said goodbye to the guests, there was no sound at all, I had completely lost my voice. Elodie took to calling me 'Darth Vader'. We laughed.

I was booked to speak at a wine conference in Bordeaux, so no voice was not ideal. Our tattooed Doctor diagnosed laryngitis and suggested rest as the first, and most important, method of recovery, along with an asthma pump to help it along. He reckoned, taking that route, I'd have my voice back in about 10 days. Stronger steroids were possible if I needed a faster recovery. Given the talk was two weeks away, I decided the slow recovery would be fine. I never realised how critical my voice was to normal life. Sean joked it was a husband's dream, no more nagging. I wrote messages on paper, but they were easier to ignore. The frustration of not being able to instantly express what I wanted was extreme. It was like a forced silent retreat.

By the time the conference on Wine, Environment and Society rolled around, my voice was husky but strong enough to participate. A professor from Hochschule Geisenheim University offered a dense presentation showing everything from carbon dioxide emissions to climate change in each of the cities that formed part of the Great Wine Capitals network. The conference had been organised by The Great Wine Capitals, a network of cities that are part of internationally renowned wine regions. It is currently made up of Adelaide in South Australia, Bilbao for Rioja in Spain, Bordeaux in France, Cape Town in South Africa, Lausanne in Switzerland, Mainz for Rheinhessen in Germany, Mendoza in Argentina, Porto in Portugal, San Francisco for the Napa Valley in the USA, Valparaìso for the Casablanca Valley in Chile and Verona in Italy.

He told the audience that vineyards were the major consumer of fungicides, more than 60 percent of EU consumption, relative to less than 5 percent of the farmed surface area, and major consumers of herbicides. He stated that winegrowers could not work without glyphosate herbicide. I wanted to jump up and shout how wrong he was. More than 10 percent of French vineyards were certified organic at the time and therefore not using glyphosate herbicide. All those growers proved him wrong.

Worldwide glyphosate had grown more than ten times in twenty years[13] despite us knowing it was bad stuff. It was forecast to keep increasing. The World Health Organisation classified glyphosate as 'probably carcinogenic in humans.' The biggest legal case against a key manufacturer of glyphosate initially awarded 289 million US Dollars to the plaintiff, who had cancer, and his family[14].

Many people would recognise the 'Roundup', the Monsanto glyphosate brand name, more easily than the word glyphosate, thanks to decades of advertising. Bayer CropScience, the company that bought Monsanto in 2018, had revenues of 20.2 billion USD in 2021. It was easy to see how an entity of this size could use financial muscle and powerful lobbying to its benefit.

There are many reasons for concern about glyphosate, particularly human health, soil health, and long-term sustainability. Some weeds have also become resistant to herbicides. At Château Feely, we had not used herbicide since 2005 when we bought the farm. After a few years, the 'bad weeds' had all but disappeared, competed out by 'good weeds' like clover. All plants come to do something, they have a purpose; they are bringing soil equilibrium back. Each wild plant tells us something about what is going on with our soil. Sean says there are no 'weeds'. All plants have value. Our words need to change so our attitude changes.

When we started gardening in Dublin the gardening magazines had a fetish against dandelions. I took it on and tried to remove them, digging them up one by one, without success, thanks to their very deep taproot. Now I love dandelions. The French name is *pissenlit* – pee in the bed – for its diuretic properties. Our English name comes from French too, *dent de lion* – tooth of the lion – for their jagged leaves. The flower buds can be brined and pickled to create a caper knockoff. I adore the leaves in a salad, eating the young flowers straight up, and the pure joy of seeing their seed heads, perfect circular puffs, each seed like a fairy that floats on the breeze.

When we arrived, like those gardening magazines, the farm adviser we inherited was obsessed with the *mauvaises herbes* – weeds – he saw in the vineyard.

Four were 'rampant' *chiendent*, 'Elymus repens' in Latin, common English name 'couch grass'; *liseron,* 'Convolvulus arvensis', common name field 'bindweed'; *chardon marie*, 'Silybum marianum', common name 'milk thistle', and *ronce*, 'Rubus fruticosus', common name 'blackberry'. I loved the last one, picking its berries in the cool morning was one of summer's great joys. When I walked the hedgerows in berry season, even if I wasn't hungry, I found myself scanning for luscious black delights.

In our first year, that farm adviser was so disgusted with the weeds that he wanted us to herbicide the vineyard. For him, the only way to beat them was to use glyphosate and a pre-emergent herbicide. Ironically, the herbicides used by our predecessors gave the 'bad weeds' the head start. We have never used herbicide, the 'good' weeds often out-compete the 'bad'. We have far more of the good guys.

It is 'eco-logic'. Ecological and logical thinking. We can't sit on sun loungers and do nothing; we would soon have a forest instead of a vineyard. We keep the wild plant growth in check by mowing between the rows and using a mechanical hoe under the vine rows. Certified organic, as we are, guarantees that no systemic chemicals or weedkillers have been used in the farming of the product.

While I disagreed with the professor on some of his statements, it was clear he thought deeply about the issues at stake. As part of a large university in Germany with a major viticulture section, he had to tread carefully. Germany's vineyards were 3 percent organic at the time, and Germany was home to two of the largest agricultural chemical companies in the world, Bayer CropScience and BASF.

In the session after the professor, the Sustainability Manager of *Vingruppen* in Sweden said we needed a 'sustainable wine' label that was wider than organic and based on the 'Better Cotton Initiative' (BCI). Consumers had become aware of chemicals used in cotton production and were demanding more certified organic cotton than was available.

The Better Cotton Initiative set up a charter to train farmers on how to better manage water, use fewer chemicals and less of them, improve soil and biodiversity, and offer better work conditions. That was a promising start. Large retailers and brands could also be members. At the time of writing this, in 2022, the BCI website outlined that if retailer and brand members committed to including 10 percent of BCI cotton in their total cotton orders, they could use BCI in their marketing and on all cotton garments, even garments that did not have BCI cotton in them. The website said there were 17 producer members and 275 retailer/ brand members.

As I read these details on the BCI website, I felt frustrated. BCI was a winner for large manufacturers and retailers, but it muddied the waters for certified organic cotton and created confusion for consumers. With organic certified, systemic pesticides and herbicides cannot be used. The input invoices are analysed. It is quantified. It is the only way for end users to be sure of no systemic pesticides for themselves, the producer, and the environment.

With BCI, growers can continue to use systemic pesticides. They are encouraged and educated on alternatives, but there is no obligation not to use systemic pesticides. In the BCI Impact Report for 2020, they showed that in Turkey, BCI farmers used 3 percent less systemic pesticides than non-BCI farmers. In India, it was better, with BCI farmers using 23 percent less systemic pesticides than non-BCI farmers. But they were still using systemic pesticides.

When we bought our 14-hectare farm with the dream of creating artisanal organic wine, advisors told us organic farming was too risky. Only one percent of the French vineyards were organic at the time. We ignored the advice and followed our sense of what was right. By the time I spoke at the conference, more than a tenth of French vineyards were organic and it was an important and growing part of the market.

At the conference, I shared why some of the farmers in our small community had decided to go organic. At one farm, the family's five-year-old daughter got leukemia and there was no history of it in the family. Their research suggested it was the systemic pesticides they were using on the farm. Another farmer's failing soil health was the catalyst. Initially, their yields had gone up with chemical farming, but after more than twenty years, yields fell despite increased doses of chemical fertiliser. The negative effects of chemical farming on the life and health of their soil, and the quality of their wines, were noticeable. Another farm converted after losing a court case to an employee who experienced nervous system disruption from exposure to a systemic pesticide.

As farmers we are responsible for the health of air, water, and soil on our farms, for us and our families, but also for clients, neighbours, and the wider environment. Over the previous fifteen years of organic farming, I had learnt that it was important to follow your intuition for what was right, rather than what looked profitable. Doing what was right for the planet provided benefits beyond expectations. Success was not about more, particularly not more at the expense of more pollution and pesticides.

As winegrowers offering wine tourism, we were in a unique position to share information about farming and the environment with clients. With the climate crisis, it was even more urgent to keep learning and sharing. As if to confirm my thoughts, we won the Gold Trophy for education and valorisation of environmental practices at the first national wine tourism awards. It was a massive win for a tiny farm in an uncelebrated appellation. Most of the other Gold Trophy winners were big organisations in famous appellations.

The prize included a trip to Japan to promote French wine tourism. It wasn't a market we particularly wanted to develop for our wine. We were too small, and it was too far away, but joining the group of wine tourism professionals on the trip was a networking opportunity. As part of an official French delegation, we were welcomed in ways I would never have been as a simple visitor.

Japan has a strong culture that is very different from the African, North American, and European cultures, I had experienced in my life. There were times when I was shocked by how much of an outsider I felt. The elevators in Tokyo's main station had floor numbers in Japanese characters only. I had no idea which button to press. Signposts were illegible since I couldn't read the alphabet.

Our agenda was packed: working meetings with the local teams of *Atout France*, the organisation responsible for promoting tourism in France, business meetings with Japanese travel agencies and tour operators, and a glittering reception at the French Embassy for around fifty wine and travel professionals and journalists.

My colleagues were Gold Trophy winners from the other categories in the competition: Château de Pennautier in the Architecture and landscapes category, Champagne Maison Ackerman in the Art and culture category, Champagne Pannier in the Business wine tourism and private events category, Cité du Champagne Collet for the regional promotion category; Bordeaux Château Vénus in the Creative initiatives and originality category and Château Guiraud Grand Cru Classé Sauternes in the Catering category; Cahors Château de Mercuès in the Accommodation category, and one winner from Burgundy, La Chablisienne in the Family vineyard category.

We were accompanied by regional leaders for Tourism in France. It was a heavy-hitting band of comrades that boasted Michelin star restaurants and aircraft flying over vineyards. I had moments of feeling 'imposter syndrome' but I was proud to be there representing Château Feely and French wine tourism.

After the intense business meetings in Tokyo, we needed relaxation. Japan was keen to increase its wine tourism and to hear how France had developed the sector. We were whisked away on a 'Shinkansen' high-speed train to Nagano prefecture, a region where wine tourism was starting to take off. Two winery visits stood out for me. The first was *Vino della Gatta Sakaki*, a garagiste – a term for a small-scale entrepreneurial winemaker – that had imaginative cat labels and a winemaker that had learnt the ropes in California. The second, *Villa d'Est*, was more French influenced as the name suggests.

The founder, a Japanese artist and essayist, studied literature at the Sorbonne in Paris in the late 1960s. The small, high-altitude vineyard, offers visits, a restaurant, a magnificent natural setting, and a boutique selling wine, art, and tableware and stationary decorated with the artist's designs. I loved the *Villa d'Est* Chardonnay with its luscious tropical fruit, white flowers and hints of butter and vanilla, but it was the view of vineyards and layers of mountains on the opposite side of the valley that stole the show.

The following day we ate from stylish Bento boxes decorated with cream and red squares, tasted local wines and watched snowy mountain peaks pass by on a luxury vintage train, the Rokumon, that took us from Karuizawa to Nagano city. En route, we stopped at Ueda for a Taiko drum performance on the station platform. Watching was invigorating, and when they offered me an opportunity to play, I discovered the act of drumming was even more energising. The vibration and the rebound of the drum give the player an intense physical experience.

Zenkoji, a Zen temple in Nagano, felt like a gentle push on the road to yoga. At the entrance, I took in smoke from the incense of purification. The temple was founded in the 7th century when Buddhism arrived in Japan. It houses Japan's oldest Buddhist statue and is said to offer all people who pray there salvation.

After walking through the main temple, visitors are invited to open the door to heaven. Somewhere down a pitch-dark ancient corridor, in the basement, is a hidden metal key or handle. Turning it is your ticket to heaven. The metal key is located directly below the sacred image of Buddha, a golden sculpture, that has been in the temple since its beginning in the year 654 hidden from view. We shuffled into pure darkness, feeling along the wall of the corridor for what felt like an eternity. A whisper of guidance from those ahead led me to a metal handle. I turned it. The key also promised an instant touch of youth. Perhaps it was the adrenaline of passing blind through an ancient temple passage, but I felt rejuvenated when I stepped in the light.

A Buddhist monk led us from the temple through a zen garden with a fountain and a wishing tree, to a peaceful hall in the outbuildings. He invited us to take a seat on the wood floor, then began a short meditation. I felt the bliss of it percolate my body and settled in, so happy to be there. Too soon, he gently called us back, then offered us the *kaisaku*. As I heard a thwack on the back of a colleague, followed by two more, I wasn't sure I wanted it.

Eyes closed, adrenaline in the belly, I took my beating. The whacks with a solid wood baton gave a minor sting, no long-term or significant pain, in fact, it felt refreshing. *Kaisaku*, a flat stick whacked across the shoulder blades or the muscle of the back, is called the 'stick of compassion'. It is used at the end of the practice, like we had experienced; or to bring attention back if a monk's focus wandered during meditation. Even in the latter case, it is not considered a punishment, but rather a compassionate way to reinvigorate, and awaken, the meditator.

Another form of invigoration, Japan's world-famous thermal baths from natural hot springs, called '*Onsen*', was waiting for us. But first, on arrival at our hotel, we were treated to a sake tasting and a twelve-course reception dinner with the local tourism players, mayors, and wine producers, that included gift-giving and excited exchanges via hand signs, mobile phone translation, and human interpreters. The traditional Japanese menu included many delicacies, miso soup, sushi, and pickled seaweed. Everything was luscious, save the cod balls, or '*shirako*', sacs of cod sperm. Ours were full sacs, white, opaque, and firm. I ate one and almost gagged. *Shirako* are believed to promote anti-aging and are full of protein and B vitamins. In the washroom, I checked my look in the mirror and decided it didn't work as well as finding the key in the corridor. Perhaps it was too much sake. Or perhaps *shirako* worked better for men.

Our interpreter had explained the *onsen* etiquette before we arrived. Even though it was public, you had to bathe nude, silence and slow movement were encouraged, and showering before entering the baths was required. After all the shots of sake, I felt ready to meet the first requirement.

In my room, with sliding wood doors, a traditional floor futon and a wood shower room, I suited up in the light kimono gown and slippers provided and followed the signs for the women's *onsen*. As I showered, I looked surreptitiously through misted glasses at what the Japanese women were doing. Groups were hanging out together whispering, others were soaking in silence. Walking slowly, nothing but a tiny towel in hand, to the big bath, I stepped in delicately. It felt soothing and peaceful. I sank into the steam and silence interspersed with gentle murmurings in Japanese.

People disappeared outside through a side door and after a while, I garnered the courage to follow. The cold air was a shock to my naked body. Three wooden spas were spaced out across the deck, one for a group, and two smaller ones for individuals. I stepped into the last, and only free one, feeling like a princess that had won the lottery. It was like a wine barrel but bigger, wide enough for me to sit in, and deep enough to cover my entire body. As I lay in the mineral-rich water, the steam rising around me, I felt a sense of well-being like the one I had felt in meditation.

The water was hot but bearable, and the air temperature was below freezing. The stars shone with a brilliance particular to icy, clear air. Slow breathing in the steam, I lay my head back on the side of the spa and felt better than I had in a long time. The laryngitis had taken a toll. The heat and minerals of the natural springs improved circulation, moisturised the skin and sped up recovery. But most importantly, they encouraged relaxation.

Back at home, the relaxation of the *onsen* baths forgotten, I was firing on all cylinders, my mind going a mile a minute. It was Friday night; Sophia had returned from her school week in Bordeaux. I was telling Sean about the Japanese group I had welcomed that day, the decisions we needed to make, and the volunteer week I had planned.

'Calm down Caro,' said Sean. 'I was all relaxed and calm preparing dinner and you're creating a tornado. Relax.'

'Mum, what about my EU care card? Did you send it to Bruno?' asked Ellie. 'I need it for my school trip to Italy.'

'Can we go to Bordeaux to watch 'Little Women'?' asked Sophia.

Relax. What is this thing they call relax? For a mother and businesswoman, it was hard to see where that could fit in.

'I don't want to drive to Bordeaux,' I said. 'It takes too much time and too much petrol.'

'Oh, I really want to go,' said Sophia.

'Yeah,' said Ellie. 'You could pick me up in Eymet after school and we could go into Bordeaux to see the film with Sophia.'

There was no way I was driving to Eymet, then to Bordeaux, at least three hours, to go and see a film and then nearly two hours home. For ecological and mental health reasons it was a no. The demand was like a mild version of Dana Carvey's clip, 'Teenagers are Nightmares', from his 'Netflix is a Joke' comedy on YouTube, where his son wants to drive 13 hours to Lake Tahoe to snowboard for 11 minutes[15]. But they were so keen. I needed a strong response to get them off the idea.

'No way!' I said. 'There's no way I'm doing all that driving. That's final. No discussion.'

Perhaps I was a little too strong.

'Chillax,' said Sophia.

Ellie looked like I had terminated the most prized adventure of her life.

'Calm down all of you,' said Sean. 'You're too edgy.'

'Maybe it's the full moon,' I said. 'At NatureElle, my eco-hair salon, someone came in full of fire and brimstone and we agreed it was the moon.'

'I think you all need to get some rest,' said Sean.

'I suppose you need some,' I said, wondering to myself if it was true. 'But you can have too much.'

At the same time, Ellie said, 'You can't have too much.'

We collapsed in laughter.

'I guess it depends on who you are,' said Sean.

There was no question I needed more of it to make sure I didn't fall back into illness.

Sean encouraged me to take a course that winter, to get away, to do something to help me find a better work-life balance. For years I had done yoga and was keen to know more. I found a month-long intensive 200-hour Yoga Teacher Training in Biarritz and knew I had found the perfect getaway for me. I loved the sea, the course was reasonably priced, and it was close enough that I could get home easily and quickly if necessary. Elodie, our apprentice, was fantastic, I could rely on her to keep the tasting room and deliveries running. The universe had aligned everything perfectly.

CHAPTER 16

Yoga in Biarritz

I took off on my yoga adventure filled with joyful anticipation, but also a little nervous. I hadn't been away from my family or our business for this long before. I was an intermittent yoga practitioner, still essentially a beginner. Who was I to think I could become a yoga teacher? I pushed aside my negative thoughts and turned up the music to enjoy the drive.

Biarritz is a jewel on France's Basque coast, a natural bay with three stunning beaches and a vibrant town centre. It is an upmarket surfer's paradise with an ecological twist. My studio apartment in Biarritz's St Charles quarter had wood floors and high ceilings. I felt instantly settled as I unpacked my bag and the box of organic food, including homegrown vegetables, prepared by Sean. It was still light, so I locked the door and skipped down the stairs for an orientation walk. The apartment was perfectly located, five minutes from the yoga training, and ten from the *Grande Plage*, the main beach.

It felt strange living on my own, with no husband or children. I made a pot of lentils and rice, enough to last a couple of days, enjoyed a glass of red wine, and read some of the recommended pre-reading. I could hear neighbours above me, walking on their wood floor, playing the clarinet, or watching television. They, and the road outside, kept me awake. I was out of practice with city life. I needed earplugs.

The next morning a sign inside the door at La Maison Rouge, the location of the yoga studio, invited me to take a training manual and climb the stairs to the top floor. A small knot of people was gathered in a meeting room, a glass wall on one side and a large window onto the garden on the other. Our trainer, Sunshine Ross, of Awake Space, welcomed us and set the tone for our training, relaxed but strict. Her slight body and wide smile belied serious strength of muscle and character.

We were a diverse mix of wannabee yoga teachers ranging in age from 23 to me. Freya, the youngest, was already a gifted yoga teacher. She wanted to combine her university studies and yoga. Marco, the only man on the course was a gentle, strong Spaniard who wanted to give up his job as a factory foreman to become a full-time children's yoga teacher. There was a vet, a chef, an IT professional, and someone who had worked in big pharma. Some had turned to yoga after experiencing burnout, some were already in the business in one way or another, and others were there to explore. Over the month I would get to know these people, and yoga, as close friends. The course ran from 7 a.m. to 7 p.m. every day except for two weekends when we had some down time.

'Before we get into the programme, I'd like to ask who is here primarily for the movement part of yoga, who is here primarily for the spiritual part, and who is here for both,' said Sunshine. 'First, those who are primarily for movement?'

I put my hand up. The only one.

'Movement and spiritual?'

Most hands went up.

'Mostly spiritual?' One hand went up.

I was not expecting a strong spiritual side to yoga. I had started doing yoga for pregnancy almost twenty years before, then picked it up again in my late forties to improve my flexibility and for gentle exercise. I had been a keen runner and a twinge in my knee forced me to change tack.

As we progressed through the first week, I learned that yoga was deeply spiritual. It offered healing to mind and body. It was a practice of union of body and soul, of consciousness and infinite. It was about balance, strength, and movement, but also about meditation and relaxation.

Yoga offered ways to look after ourselves, to be better to ourselves, so we could be better to others, and to our home, planet Earth. Using breathing exercises, movement, and care for our spirit or soul, we were better able to care for others. By controlling our breath, we could control our minds and our emotions.

I learnt that breathing, movement, and meditation, represented three of the eight 'limbs' of yoga. *Yama* was external discipline including '*ahimsa*' not harming, truthfulness, not stealing, impeccable conduct and loyalty, and not being greedy. *Niyama* was internal discipline including purity, contentment, intensity of discipline, self-study, and mindfulness. *Âsana* were postures, the movement part of yoga. *Prânâyâma* was breath regulation. *Pratyâhâra* was withdrawal of the senses. A simple example was closing your eyes and looking inwards during a posture. *Dhâranâ* was concentration, an example was concentrating on one spot to keep your balance in a pose. *Dhyana* was meditation. The destination of all of this was *Samâdhaya*. It was oneness, integration with the universe. By meditating and following the other seven limbs, anyone could reach this state of ultimate bliss.

Each day we started with a half-hour of meditation and breathing practice, then a two-hour yoga class followed by two hours of theory. After lunch, two hours of theory preceded two hours of yoga and practical teaching. Being a yoga teacher was a responsibility, to share the message to love and respect your body, to give it the required rest, care, and movement. Its message was made for me. I hadn't done that in the previous few years, and I had paid the price. Now I was learning how to do it, and how to help others to do it too.

As I travelled home for the halfway weekend, I tuned into French national radio. The United Nations Conference of Parties on Climate Change, COP25, was playing out in Madrid. A journalist attending the conference gave her impressions to the talk show host. She summarised the conference as a cop-out, a disaster, because large players like the USA, Brazil, and Australia, countered every proposal. It was seriously bad news. I had an inkling how bad, perhaps more than most people listening. My jaws clenched and my body tensed.

The show host commented; 'there is nothing we can do as individuals; we need governments to act'. The journalist at COP (out) 25 continued as if she had not heard what the host said. I felt my body tighten at the false message being passed to the audience, a message of powerlessness, of hopelessness, of our inability to make a difference. The host was wrong. If governments were not acting, it was even more imperative for us to act as individuals and to demand, and make, the changes necessary to avoid total climate chaos and destruction of our home. Our demands would force governments to act. We needed individual and collective action. Giving up was not an option.

I turned to the yoga breathing technique learnt that week, used by Navy Seals and first response emergency units to keep calm. It is called 'box breathing' or 'square breathing' because you can think of each set of four as a side of a square. Breathe in for four counts, hold your breath for four counts, breathe out for four counts and hold your breath for four counts. I continued with that method until I was in a calmer state, more in control of my instinctive body.

What are democratic governments, if not a set of us? What we do guides what they do. What we do guides how much, and what kind of food, energy, shelter, and entertainment, is created. If we don't buy it, no one will make it. If we don't vote for them, they won't be in power. I found great solace in the impact that one teenager, Greta Thunberg, had made. One individual. When we act, we also influence others to act, as she had done. We never act in a vacuum.

My bedside reading that year had nudged into stressful territory, 'This Changes Everything: Capitalism vs. The

Climate', by Naomi Klein, and 'The Uninhabitable Earth: Life After Warming' by David Wallace-Wells. They shocked me awake to the full-on emergency that the environmental crisis and climate crisis represented. It wasn't only the polar bears' home and their survival that was at risk. It was ours.

At that moment, more than half of the carbon dioxide released into the atmosphere by burning fossil fuels – our cars, heating, food miles – had come in three decades from 1990 to 2020. This meant that we had done as much damage to the potential for our planet to remain habitable for humans since my final year in university, as was done in all time since the beginning of industrialisation. The United Nations (UN) created a climate change framework in 1992. The world knew, but the message was not shared. It was hushed by lobbyists; swept under the carpet by climate change deniers.

We didn't act. We accelerated the harm thanks to fossil fuel companies' massive investments in lobbying and misleading reports. The world breached the safe level of carbon dioxide in the atmosphere, considered to be 350 ppm in 1986, my final year of school. By the time the UN created the climate change framework in 1992, the level was 360 ppm.

With business as usual, we were speeding towards a 4-degree Celsius increase by 2100, about one lifetime away. That would make large parts of the planet uninhabitable due to rising seas, extreme heat and more. Already climate change was taking its toll, creating a long list of devastating effects including, according to 'The Uninhabitable Earth': heat death, hunger, drowning, unbreathable air, wildfire, disasters no longer natural like outsize tornados and hurricanes, lack of fresh water, dying oceans from acidification, plagues of insects and diseases, economic collapse, climate conflict, and the multiplier effect of these acting together.

The climate crisis and the environmental crisis were not distant crises to put on the long finger (an Irish term for postponing something for a long time). We needed to look them in the face to have a hope of getting out alive ourselves and as a species.

I resolved to manage the fear but also harness it for action. My action, individual action, would lead to team action. I reviewed articles on individual action and then used the guidelines to assess our situation. I looked at our transport, food, housing, clothing, and finance. A lighter planet footprint could be better for my bank balance and my happiness too.

The website 'Treehugger' reminded me of things I could do better. For example, it was easier and more efficient to heat my body with extra undergarments than to heat an entire room. I added layers without going on a shopping spree, some thermals at the back of the cupboard from years ago, an extra outer layer, my favourite coat purchased for ten euro at a thrift shop. Another quick hit was changing our electricity supply to 100 percent green energy. I did it for a small increase in cost. It was worth it for the peace of mind that we were supporting green energy, not coal, gas, or nuclear. It encouraged the installation of more renewable energy. In France, it was easy to find details of how to change your energy supplier and to do it.

While nuclear power doesn't generate carbon dioxide, it creates nuclear waste which remains dangerously radioactive for thousands of years. It is a ticking bomb for future generations.

Treehugger also got me thinking about small changes that could make a difference to our energy consumption, like making sure our devices like the television and computer were off at night, not in sleep mode, putting my mobile on power-save or airplane mode whenever possible, to save battery and hence energy consumption, and thinking about what we watch on streaming and the weight of the digitals we share.

These 'small' changes helped offset the increase in cost from changing to a green energy supplier. Yes, my family was sick of me saying 'Turn off the lights', 'Turn off the device', but eventually it would become a habit. It didn't seem a big saving if one person turned off a light at night, but if every household in France did it, that one light per night would represent 120,000 tonnes of carbon dioxide per year. In some countries, it would be significantly more since France had low carbon dioxide emissions for electricity because of the high reliance on nuclear power.

Probably what hit me most in my evaluation, was the information about investing. I realised how much of a direct impact money had, from what we buy to what we invest in. Not supporting financial institutions that directly invested in companies contributing to the climate crisis appeared to be a no-brainer. However, moving bank and pension was not as easy as moving to a green energy supplier. Change was hard. I talked to my school friend Lee, who was back in Australia, after finishing her MBA in Copenhagen.

'I agree Caro. I want to move my banking and pension to an institution that's ethical, environmental, and not investing in the climate crisis, but it's difficult. There are few options and moving is costly and risky. I don't want to risk my retirement funds on an institution that isn't rock solid.'

She paused.

'I mean, I've worked for a mining company almost my entire career. The ecological damage they've done is probably way worse than leaving my savings with a bank that's still investing in coal,' said Lee, snorting with self-deprecating laughter. 'I feel cognitive dissonance. But there are no other jobs here. I've looked, especially since I got my MBA, but what can I say? I live in a petrostate.'

'I know what you mean,' I replied. 'My yoga training threw up the idea of right livelihood. It's essentially, 'do no harm' in your work. '*Ahimsa*', 'do no harm', is the first of the 'eight limbs' of yoga. Wine can harm if taken to excess, but in moderation, it can add to moments of conviviality and sharing. It's not simple.'

'Darn right,' said Lee. 'But God, I wish I could find something that matched my ethos. I ride my bike everywhere; I buy organic wine, then I go and work in f'ing mining.'

We laughed at the complexities of our lives and how we were dealing with the cognitive dissonance we felt. At least we were making changes and conscious of the situation. In the books and articles about decreasing my individual planet footprint and carbon dioxide emissions, two areas came up again, and again, choice of transport and food/ consumption.

Number one was to avoid the car, choose to walk or bike or take public transport. Years before, I had created an incentive for our daughters to ride their bikes, or walk, to, and from, primary school. I calculated exactly what we saved in a month by not driving, and paid them that difference. It was a win-win, as many ecological changes can be.

My decision tree for using the car became stricter. We regrouped things around one trip to our local town of Bergerac. When I travelled to an event, I sought out colleagues to do '*co-voiturage*', shared transport. It meant I got to know new people or had time to chat with friends.

Another major area of transport was air travel. The best option was to choose another form of transport, particularly for short haul. But if we had to fly, I resolved we would offset our carbon dioxide by planting trees. It didn't make it right, but it eased my dissonance. Some airlines offered the option to buy offsets when you bought your ticket. I also found online services that offered ways to offset if the airline didn't provide it. But planting your own trees, if you have space, is better, or finding a local project for renewables to contribute directly to, this way, you are sure that you are making a difference.

Coldplay announced their music tours were on hold while they considered how to make their concerts more environmentally friendly both in terms of travel and waste. I loved their music, and their initiative made me love them even more. Vogue Italy ran a series of artwork instead of doing fashion shoots to avoid the massive consumption of air travel and the single-use waste generated by an international fashion shoot. They showed that big names and big brands could help change the world.

Second after the car, at the individual level, was food. Eating less meat and dairy was a massive step. We were lucky to live on a farm where Sean grew most of our vegetables. We ate a small amount of animal protein, eggs from our chickens and meat from local organic farmers that we knew, with their animals on pasture all year round. While I lived in Biarritz, I ate exclusively vegetarian.

Growing your own is the perfect solution. No carbon dioxide is used to get the food to your plate, and the plants are part of the solution, taking in carbon dioxide as part of their cycle. When we were city professionals, I loved gardening. Like yoga, it offered a great way to be active and to relax, with the big benefit of good food as a result. At our farm in Saussignac Sean did all the gardening, but I was keen to get back into it.

Buying organic food supports farmland that holds carbon dioxide via regenerative agriculture rather than releasing it. Supporting local rather than food flown in from the other side of the world is also a great way to connect to community. I loved the Bergerac market and the organic food producers there.

We are not powerless. If anything, we are more informed and more empowered to change the world than ever before given the ease of sharing enabled by the internet and digitisation of books. When we act as an individual, we motivate others to do the same. We are all in this together.

While personal action is good, we also need to create systemic change. One person changing their bank is not enough to change the bank's direction, but a movement of many people collaborating to do it at the same time might do it. The founder of 350.org Bill McKibben started an over-60s movement called 'Third Act' that created an event in 2023 to raise awareness of bank's complicity in the climate crisis. Another movement by students created enough pressure for large institutions, like Harvard University, to divest from fossil fuels. Removing money funding fossil fuel expansion stops it in its tracks. I realised that speaking out and using my vote to create change were as important as my personal actions.

At the time of Cop-out 25, and my stack of scary books like 'Uninhabitable Earth', yoga saved me from spinning into a TSA – total stress attack – a handy term Sophia had coined for when her mom freaked out. Since then, I have discovered books that offer a gentler and more positive approach to dealing with the climate crisis. 'Zen and the Art of Saving the Planet' by Thich Nhat Hanh, he of sage relationship advice, is a must-read. His book on Love included a foretaste of this. In it, he recommended using the mantra 'I know you are there, and I am happy about it', beyond your loved one, for example on the full moon, plum blossoms, an oak tree. In that moment of recognition, we become aware of the miracles in the everyday around us. I said this mantra to cherry blossoms recently and felt a spiritual connection to them, a sense of interbeing that filled me with bliss.

In 'Zen and the Art of Saving the Planet' Thich Nhat Hanh offers hard-hitting messages, 'if we continue the way we are living, the end of our civilisation will be certain', followed by 'Only by waking up to this truth... will we have the insight and energy' necessary to change. The discovery of this work and others that invited me to see the climate crisis in another way was still to come.

The faster I informed myself, the more cognitive dissonance I experienced. It felt like we were living in a contradiction. It was hard to circle the environmental, spiritual, and social aspects I wanted in my world with the capitalist economic system we were bound to. I had grown up with the idea that success was a big house, a big car, a big bank balance and fancy holidays. Social media and advertisements reinforced those ideas incessantly. I had to rewind the tape that had been playing since the beginning of industrialisation, the one that said 'more, more, more' and replace it with one that said, 'less is more.'

William Blake (1757–1827), English poet, thinker, and artist, said, 'You only know what is enough when you have had too much.' He also said 'No bird soars too high if he soars with his own wings'.

There were so many messages for the climate crisis in these two quotes. Had we humans realised we had had too much? Given we had soared on other's wings, metaphorically, using fossil fuels from another era, had we soared too high to save ourselves?

In a culture of more, it's difficult to sell the idea of less.

Henry David Thoreau (1817 –1862), an American naturalist, essayist, poet, and philosopher, best known for his book Walden, a reflection on simple living in nature, wrote, 'A man is rich in proportion to the number of things he can let alone.'

In yoga, this was called '*vairagya*', letting go of the unnecessary. But what we resist persists, so it is not about fighting to say goodbye to something, it is about resting on the positive. Yoga helped me manage the stress of the climate crisis. I knew it could also contribute to creating a society with more spiritual happiness, a deeper happiness than that promised by new shoes, a new phone, or a fancy holiday. I could see so much potential in it for health, joy, and the future of our planetary home.

In November 2019, around the time of COP-out-25, the monthly average carbon dioxide in the atmosphere recorded at Mauna Loa Observatory in Hawaii increased to 410.48 ppm. We were addicts that could not stop, aided by lobbyists intent on unstitching what small progress was being made.

I returned to Biarritz on Sunday. After the long drive, I walked to the sea to stretch my legs. Halfway between the yoga studio and my apartment was a tiny organic shop with local producers and bulk bins that included favourites like lentils and dried figs. It was my preferred food shop. For lunches out, an organic café 'Nuts', was my go-to destination, and became a client for Feely wines. Their fresh juices were like a shot of cosmic energy.

Freya the young yoga guru, Cora the vet, and I had grooved a habit of taking a swim at the *Plage du Vieux Port* every couple of days. It was a small beach, a little larger than 'our beach' in Vernazza, but with the same feeling, protected by a head and surrounded by breathtaking views. That Sunday afternoon I texted my collaborators, but they had other plans.

Surrounded by the local 'polar bears', members of an association of swimmers, '*Les Ours Blancs*', that braved the sea all year round, I swam out further than I usually did. From there, I could see up the coast to Spain, hills dropping to sea in lilac layers. I floated on the rollers, barely feeling the cold. A few months later Sean and I would return and swim here together creating a Vernazza Two. We were falling in love again.

We were asked to find a sentence that spoke to us in the Yoga Sutras. My selection was:

'Contentment brings unsurpassed joy.'

The bliss I felt in meditation was mind-blowing. Yoga was far more than movement. One morning before dawn, our class took a meditation walk from the studio to the *Plage de la Côte des Basques*. We walked in a long slow line, a few metres apart, down *Avenue Reine Victoria* to the *Grand Plage*, then up the seawall and down to the *Porte des Pecheurs* passing *Chez Albert*, where Sean and I would enjoy a romantic dinner as we used to in Dublin.

Putting one foot slowly in front of the other, the figures around me vague in sea mist and night lights, I entered a deep meditation drifting in semi-conscious bliss. We rounded the sea cliffs past the walkway to the Virgin's Rock then circled around our beloved *Plage du Vieux Port* and down to a concrete area in front of the *Plage de la Cote des Basques*. I could have kept walking, but Sunshine invited us to do our own private one-hour yoga class while she sat on a mat and looked out to sea.

She explained that she was not allowed to give yoga classes in a public place without authorisation, contravention would incur a hefty fine. It was intel worth knowing. I guessed it was about insurance and responsibility, but it seemed a pity. For a start-up yoga teacher in a big city where rents would be high, outdoor classes could offer a great starting point.

'Why did you do this yoga course?' asked Mum.

'To teach yoga,' I replied.

When I arrived, the objective was to teach yoga movement. Over the days and weeks, I realised how important the spiritual side was. The meditation and the invitation to relax, properly relax, were sent to save me from burnout. As we shared our spiritual journeys and some of the other participants described their burnouts, I realised how dangerously close I had come.

Meditating quietly on the mat every morning gave me a stability of spirit and a sense of internal peace. By the end of the course, my body was fitter and stronger, and my mind was relaxed. I felt like the yoga training had given me the tools to handle anything that came at me. I had breathing techniques like *Samavritti Pranayama*, box breathing, to stay calm in a crisis. My soul was nourished with spiritual practice and meditation. I felt like this course was my destiny. I didn't know how much I would need it. The world as we knew it was about to grind to a halt.

CHAPTER 17

Bottling in the time of Corona

The speed at which the Covid-19 virus changed our lives was shocking. At first, I discounted the story since recent disease scares had come to nothing. Then it crept closer, to Italy, to France, and wham, the world was in lockdown and cancellations were flooding in as people's holiday travel was revoked.

I processed thousands of euros of refunds and prayed for hospital workers and people directly affected by the virus. At night my mind churned. How would we cope as a business dependent on tourism? What about the wine business? Would we be able to get supplies like glass and cork? To confirm my worries the news announced bottle shortages, partly due to factory closures, and partly due to border closures. Even without a pandemic, our annual bottling gave me nightmares.

Our suppliers confirmed our orders for bottles, corks, labels, and capsules. Then the corks were lost in transit, sucked into a Coronavirus vortex. We had booked full-service bottling so changing the date or cancelling would incur significant fees. A new cork order would not be completed in time.

I walked outside to gain perspective. The grass was soft, the sky was blue. The planet was turning. I lay down and was enveloped in the fragrances of eucalyptus and forest floor. I felt free like I was floating up into the glorious immense blue, unbroken by flight paths. The pandemic had created empty skies and empty roads. As I lay on the ground, I took in the earth, how it felt, how it smelt. A nature meditation. Bliss enveloped me. A couple of days before the same grass had been rich with floral scents. At that time, we had been in the middle of a flower zone on the biodynamic calendar, now we were in a root zone.

The biodynamic calendar plots the earth relative to our moon, the planets of our solar system, and the twelve constellations that surround us. It helps us to plan our farm and winery work. All farmers used a calendar like this, called the farmer's almanac, until the 1970s when chemical farming took over. We can all see the direct effect the sun and the seasons have on our lives. A little less obvious, but still visible, is the moon, and particularly its effect on the tides.

In the calendar, days are categorised for how good they are for planting, or harvesting, based on what is going on in the sky. We can even find zones that are more propitious to certain types of plants than others. The calendar connects each of the four primary elements earth, water, air, and fire, with a part of a plant, earth with the root, water with the leaf, air with the flower, and fire with the fruit. If the plant you plan to cultivate is for the root, like a carrot, then planting on a root day in the planting zone of the calendar, would be ideal.

Even the taste of wine is affected by the day. On fruit days, fruit aromas and flavours are more obvious. On flower days, floral notes are more obvious, things like orange blossom on Semillon, elderflower on Sauvignon Blanc, or violets on red wines. On leaf days, herbaceous notes are reinforced. For example, garrigue herbs or mint on our Semillon; the grassy, asparagus elements of our Sauvignon Blanc and fennel or mint on Feely reds. On root days tannins and earthy notes are more obvious. We aren't the only ones that have gone lunar, some big retailers only organise their professional tastings on fruit days.

Refreshed, my lungs filled with earth aromas, I returned to the office, the limestone crunching underfoot. It was a perigee; the moon was closest to the earth on its orbital cycle. All was quiet, the quietest it had been in years. There were no cars roaring on the road below the farm. I felt revitalised, reset, and ready to face my challenges.

French schools announced that they would close for two weeks. The two weeks would stretch into around six months. As I dealt with Covid business challenges like lost corks and cancellations coming down like confetti, our daughters adapted to online school. They needed their digital devices for social exchange but also for teleworking. But they were staying up too late, and then getting up late, like in the holidays. To combat the trend, Sean and I declared two new rules. Devices had to be off by 9.30 p.m. and the day needed to start by 8 a.m. like a school day. We wanted everyone up at a reasonable time and keeping to a schedule, so we didn't dissolve into disarray.

Sean kept steady, working in the vines, the winery, and the garden, and making us healthy dinners at night. I fielded cancellations, processed refunds, and sought creative ways to deal with the crisis. Orders flowed in after I sent an email saying we could still ship wine. Then our local post office, our primary method of shipping wine, announced that, due to the pandemic, they were closing until further notice.

Mine were meagre worries compared to those working directly with the virus, suffering from it, or worse, who had lost loved ones to it, but my mind kept racing to the future. There was deep stress in the crisis even though we weren't on the frontline, and we were healthy. I checked in with my body and found my muscles were tensed as if I was preparing for fight or flight. I did a minute of box breathing.

That night Ellie told us she was not happy with the 9.30 p.m. device restriction. It showed we didn't trust her. We debated the situation, considering all sorts of solutions and alternatives. Eventually, Sean proposed phones down at 9.30 p.m., they could keep their laptops, and the internet would be switched off at around 11 p.m. when he finished watching his show. Whenever Sean turned off the internet, I struggled to get it on in the morning and wasted hours of potential work time. With the pandemic, the cancellations, the shipping hitch, and the lost corks, the idea sent me off the deep end. I cracked.

'Get a grip!' I yelled. 'Think about what really matters. The world is experiencing an unprecedented crisis and here you are fighting about access to your phone. What the hell?'

I shouted, then I cried and shouted. It was the wildest and craziest I had ever been. I melted down. So much for yoga zen. I ran to the bathroom to wash my face and regain some composure. When I returned, we discussed the situation calmly. We devised some modifications. They could keep their phones, but everyone had to be in the kitchen for breakfast by 7.15 a.m. on weekdays and 9 a.m. on weekends. We were obliged to do an hour of gardening with Sean every afternoon, and some other physical activity, running, gym, or yoga, for at least half an hour every day.

Our daughters were fighting for their independence, a normal process for teenagers. Their phones were their source of interaction, their link to the outside world in this strange new lockdown realm. I needed to see their side of the story.

The following day I did a stress-buster yoga routine and felt better. In the search, I found a sequence to boost immunity. Ellie and Sophia joined me in the Wine Lodge living space that had become our de facto gym. Furniture was pushed aside, and mats, blocks, weights and skipping ropes, had taken their place.

I had forgotten how beautiful it was in there, oak and poplar panelling framing massive glass windows that looked onto the old Semillon vines and up the valley to Saussignac. Ellie turned on some music and Sophia led us in a workout. I recognised the music and started singing 'My feet below the water.'

'Those aren't the words,' said Sophia, giving me a look. 'It's 'My face above the water.'

We lay back on our mats and roared with laughter.

Everything was upside down.

On the work front, the post office had closed, but its absence led me to another courier that was still operating in our area. When they collected the boxes, I opened the barn doors so the driver wouldn't touch door handles. We kept a wide berth. The driver left the paperwork on a pallet and said I shouldn't touch it for 24 hours. We shared smiles and namastes at a distance.

We were far from our extended families, unable to offer help to our aging parents, and that added to the stress of the new pandemic world. On the positive side, pollution had dropped, the sky was clear, and our small family was spending more time together, and laughing more than we had in a long time, despite my outburst. Wildlife thrived. From the office window I watched a pine marten cross the courtyard and then swing into the trees like a monkey, a couple of days later a fox sauntered across, then a deer, and later a small wild boar went snout to snout with Dora our dog, then took off when she barked.

In lockdown France, you couldn't go out without an authorisation form, authorised by... Yourself. The police were out in force to control people's personally authorised authorisation forms. Sean was stopped in Gardonne, five kilometres away, on his way to collect animal food at *Pole Vert*, a farm supply store. *Pole Vert* had never offered online shopping, but lockdown was shoving rural France rudely into the 21st century. Sean had his signed authorisation, but it was the old form, not the recently updated version. The officer wasn't happy, then he noticed the car insurance was out of date. Sean called me.

'We've paid for it. Maybe you didn't put the updated tag in the car,' I said.

'Then it must be in my office basket,' said Sean.

While he held, I searched the wicker basket on the top shelf of an antique pot stand that served as his 'office'. I found it midway down.

'I'll bring it straight away,' I said.

Sean told the officer, who said not to worry; Sean could go on good faith. I reversed back up the driveway, thanking the sky for good cops. That afternoon Sean went to Bergerac for urgent basics. He was stopped by a different cop at La Ferrière, a hamlet about the same distance from us as Gardonne, but east rather than north. The *gendarmes* had us surrounded. Sean had the insurance, and he had the new form, but he was confused about which box to tick for what he was doing, so no box was ticked. This time he got the bad cop, a guy that told him off violently and threatened a fine.

Sean recounted his experience to us that evening.

'He was so aggressive. As I drove away, I kicked myself that I didn't think to do a fake cough to encourage him to back off.'

We laughed. A tell-tale symptom of the virus was a dry cough. People looked at you suspiciously and gave you a wide berth if you coughed in public.

'I saw a video of a guy stopped at one of those roadblocks. He felt so harassed he bit the police officer,' said Sophia.

'That's what I felt like doing,' replied Sean. 'Shopping was strange too. There was an air of suspicion everywhere. The shop had a sign up limiting the amount of toilet paper you could buy. The people packing shelves and working the checkout were stressed.'

'Crazy,' I said, so happy it wasn't me doing the shopping. Since Sean had taken over the cooking, he had also taken over household shopping. He was my hero.

'The best place was *Saveurs de Saison*. There were fewer people, and the ambiance was kind,' said Sean.

Saveurs de Saison was a farm shop in Lamonzie St Martin, about ten minutes from Saussignac. It was the place where we loaded up on flour from Gilles, of *Ferme du Bayle*, a farmer I first met at the Bergerac Organic Fair a decade before; organic eggs, if our chickens weren't producing enough; organic butter and cheese.

We appreciated everything so deeply during that first lockdown. Simple homemade soda bread with a slice of Montastruc cheese took on a sense of the divine. Our personal life changed dramatically, as did our professional life.

Our new courier was overloaded and skipped two planned collections. The boxes were piling up and shipping wine direct was our economic lifeline. We needed them to keep doing what they did. After two missed collections, we had about twenty boxes lined up. Our usual man, surly at best, arrived.

'*Il y a tout ca*? All of this?' he asked aggressively.

I wanted to shout, 'It's your fault! You missed two days.' Instead, I smiled and nodded, keeping my two-metre distance. He picked up the first box with a harrumph and hit his head on the door as he went out. Karma.

The next time he collected boxes I offered the grumpy man a bottle of wine. He was more even-tempered after that, an endorsement of the calming effect of Feely wine.

I had to learn to respond with grace to all situations. My outburst about the phone curfew showed I had work to do.

That evening, as I finished making chickpea and homegrown spinach stew to give Sean a cooking break, Ellie and Sophia, chatted beside me in 'franglish', a fast-paced French well sprinkled with slang and the odd English word or phrase. At times Sean and I felt like they spoke a foreign language, something other than the two classics of the house, our native English and adopted French, grafted on when we moved more than a decade before. Sometimes we felt like they were from another planet.

Ellie had been writing on her hand and wrist, making notes of songs Sophia had played as they unpacked the dishwasher and set the table. Her hand looked like it was tattooed. I found myself thinking, 'I hope she doesn't think that looks cool and decide to have tattoos up her arm'.

As if reading my mind, she turned to Sophia and said: 'Do you think you'll get a tattoo?'

'Maybe,' replied Sophia.

Sean walked in from closing the greenhouse for the night.

'Did you hear that?' said Ellie. 'Sophia's going to get a tattoo.'

Sean's eyes went wide, then he laughed.

'I heard the run-up as I was taking my boots off outside,' he said.

'*Mais bon*, a discreet one isn't going to damage your job prospects, is it?' said Ellie, referencing my bruhaha about not trusting a doctor with tattoos. Pointing to her heart centre as a potential location she added, 'No one's going to see it.'

'You never know,' I said.

'Okay, you don't know. You're right. You might want to be a stripper and that little tattoo could stop you,' said Ellie.

'*Touché*,' I replied, laughing.

I went back to the office after dinner feeling like a rabbit in the headlights, not sure what to do about the Coronavirus tsunami of cancellations. I was about to shut down, when an email from Kelsey Toner, a Canadian I had followed for years, arrived in my mailbox. He ran an academy for tour businesses and tour guides called 'Be a Better Guide'. His free material had been useful over the years. I trusted him. This email was for an 'Online Tour Business' course, how to plan it, do it, and get it out there. I signed up.

I had been thinking about offering online wine courses for years, but I had a fear of filming myself and was unsure about how to create the courses. I had developed an outline document with notes for the different courses we could offer after brainstorming with Foo in Paris. Then the growth and demands in our physical business took my eye off the idea.

Kelsey and his team were working round the clock to create the modules, moving quickly to reshape their world for the new reality. Their target businesses, tour guides and tour operators, were some of the worst hit by Covid. The timing was perfect. Between worrying about a frost warning and the lost corks, I followed the virtual tour course modules as they were released. It was riveting and kept my mind off the chaos. Ideas began to form around new virtual products we could offer.

The bottles arrived, the front labels arrived, and the printing of the back labels that we did ourselves went smoothly, but there was still no sign of the corks.

Sean scythed the spring grass in swathes, then we raked it into mounds and forked them into the wheelbarrow and then onto the compost heap. We'd been doing it for a week and the routine was well established. I loved it, the smell of the fresh-cut grass, and the rhythmic action, were soothing. Sophia and Ellie were less keen.

'This afternoon you can do more grass raking,' said Sean pointing out the window as we ate lunch. 'I've cut a new area.'

'I saw it,' said Ellie deadpan. 'As I was having my coffee this morning, I saw my future.'

That evening Sean's homemade pizzas and homegrown lettuce were served at the table outside the winery. The glow of the setting sun lit up the courtyard. Below us, vineyards descended in lime green rows to the forest we called 'where the wild things are', our rewilding project. Deep inside was the largest and probably oldest tree on Feely farm, thought to be a black poplar. Its diameter was about two metres and it reached far above the rest of the forest. Beyond our vineyards, the Dordogne Valley was a carpet of vines, orchards, grains, and open fields. The table had become our preferred location for eating lunch and dinner since the pandemic hit. We usually didn't use the space as it would disrupt the privacy of guests in The Wine Cottage. With no guests, we were appreciating our property in new ways.

The flour of the pizzas was from Gilles, the tomato sauce was homegrown as were the onions. One had homegrown sweet potato and pumpkin, with chorizo and feta, and the other had 3 cheeses and herbs. Sean was a master gardener and chef.

'You have to respect your parents,' said Ellie. 'Especially if they make such great pizza.'

We all laughed.

'Little kids don't respect their parents, sometimes,' said Sean. 'Sometimes big kids too.'

'Parents have to respect their kids,' said Ellie, making an oblique reference to the phone access debacle.

'We try to,' said Sean.

'We try to too,' said Ellie.

Ellie wanted to go to Lycée Magendie in Bordeaux to follow the BAC Option International programme that Sophia was on. She was ready to move on to a bigger pasture. She had attended a Zoom interview in lieu of the entrance exam, which was cancelled due to lockdown. It was tough, but she was hopeful.

Sophia threw a tiny lettuce stem back into the empty salad bowl. Sean leaned in and grabbed it. I thought he was going to eat it even though he didn't like stems. I wanted it.

'I'll eat that,' I said as he threw it over the railing into the vines.

'Uh oh. Too late,' he said. 'It was the bitter stem.'

'The best part. I love the bitter stem,' I said. We all cracked up; we were opposites in so many ways.

I reached for one more piece of pizza.

'I feel like Michael Rozen, just one more piece.'

Michael Rozen, a British children's author, poet, political columnist, and broadcaster was part of our shared family humour. His 'chocolate cake' poem[16] was a favourite.

The roses on the table smelt sweet, damask with a touch of lemon curd, pure aromatherapy. Sean's parents gave us the rose bush for our first harvest. That evening I walked out with Dora, our dog, to close the chicken house. The full moon was bright, the skies were clear, the silence dramatic. On the one hand I was peaceful and calm, connected to nature, thanks to no mechanical noise, but on the other, I was scared. The uncertainty and the disappearance of 70 percent of our revenue from accommodation, tours and direct sales, and the lost corks, made me anxious.

The next day the corks reappeared at the factory where they had been manufactured. The warehouse they were booked to pass through in Bordeaux had been closed due to the pandemic, so they were returned to sender. The factory placed them on express delivery back to us, but it was touch-and-go whether they would arrive in time.

Our local supplier ordered blank corks as a backup, but I didn't want to use unbranded corks. I was in the process of resigning myself to that fate when, five business hours before bottling was due to start, a delivery truck pulled into the courtyard with the well-travelled corks. I felt like kissing the driver, but I couldn't even shake his hand. I waved excitedly from the far end of the courtyard, pointing to where he could place the corks on the winery concrete curtain and then gestured a vigorous thank you.

We had booked a full-service bottling team long before the pandemic to protect Sean's back. It was a lucky break.

Lockdown regulations made it impossible to work with a mixed team of friends, family, and seasonal workers, as we had in the past. The new machine also injected nitrogen into air space in the neck of the bottles offering better conditioning for our low and no-sulphite wines.

The crew arrived early. They were professional and familiar with working with the lockdown restrictions. But it's rare that a bottling day is totally hitch fee and halfway through the vat of Feely Générosité barrel-aged Semillon, a part on the machine broke. The maintenance van arrived in less than half an hour and quickly got the bottling back underway. It was strange for us to be in the background making sure they had everything they needed rather than on the frontline packing boxes and stacking pallets. I missed the camaraderie with friends and family, and the shared lunch, but not the aching wrists.

→→≫ ≪←←

The risk of frost passed, the trellises filled with foliage, and the vines reached skywards, their graceful canes and spiralling tendrils dancing in the breeze. We were soldered to the seasons, feeling them, flowing with them. The flowering started. Each tiny, delicate, cream flower in the right conditions would become a grape. The stamens, the male parts, five in a circle around the stigma, the female part, were doing their best to drop pollen onto her, essential to fertilise the eggs that would become the seeds around which the fruit would form. The flowers exuded aromas of honeysuckle, hops, and the hint of the smell of a newborn. It was a moment of magic and grace.

While the vines flowered, Elodie and I worked to create virtual products. I filmed solo with a selfie stick and then sent the files to Elodie for editing. She worked remotely, locked down at her parent's place, a couple of hours' drive away. The speed of rural internet pushed us to the edge, but we kept our heads and made our deadlines. Before lockdown, we had tested teleworking. Working from home saved an hour and a half of Elodie's day, and more than 11 kilograms of carbon dioxide.

Doing the calculations made me aware of how small things add up. IPCC Research shows that 2.5 metric tonnes of carbon dioxide emissions is the maximum average annual emissions per capita for us to keep within warming of 1.5 degrees Celsius by 2030 (to be precise it is 3.2 metric tonnes less an estimated 0.9 metric tonnes for government services and investments for the individual). That works out to 6.85 kilograms of carbon dioxide per day. Elodie's commute of 45 kilometres in a fossil fuel vehicle used more than that. It's easy to see that to live a 1.5 degrees Celsius life we need to stop commuting in cars, but for those of us in areas not served by public transport, and without a bicycle infrastructure, that's not easy. Treehugger's Lloyd Alter, wrote a book 'Living the 1.5 Degree Lifestyle', which outlines how to make the necessary changes. Teleworking is a potential solution. Lockdown helped us see that teleworking was viable for longer periods.

Over four intense weeks, we created a Virtual Discovery Wine Course that included 8 modules, 20 videos, documents, 12 wines, detailed tastings of them, and live online events. Like Kelsey's virtual course we were following, we rolled out the course content in real-time as we completed each module. It was a high-speed race.

We created canned modules but also live sessions. I learned that things were different online compared to the real world. The technology needed to be spot on, the pace perfect, and the presenter's energy higher to get the same effect. It was more complicated to create good interaction than in real life, but I was surprised how good online experiences could feel, like we were in the same room, not separated by many miles.

The magic came from a sense of togetherness and interaction, rather than a talking head. During our virtual Rosé Soiree for wine club and discovery course members, I used breakout rooms for the first time and found they added an invaluable inter-audience activity. Dropping into the breakouts I felt like a character in Harry Potter using magic to arrive via the chimney.

As we powered forward with our virtual tours and courses, the murder of George Floyd started a roll of protests across the USA. I felt horror seeing a snippet of the video. I couldn't watch it. When I saw violence in a film that I knew was imaginary, I closed my eyes and covered my ears. This was real.

Was there a way to stop it? Yes, by never letting it go unchecked, by standing up to it. There was no 'I'm not a racist' there was only 'I'm an anti-racist'. Silence enabled violence, as I had learned from my childhood in South Africa. A while before, Vanessa Nakate, a Ugandan climate activist, was part of a youth climate change delegation to Davos, that included Greta Thunberg and three other prominent young, white, female activists. In their coverage, Associated Press cut Vanessa from the photo. Vanessa was outraged and her video condemning it went viral. Africa generated the least carbon dioxide of all the continents, and it was most affected by climate change. In all subsequent panels and photo shoots, the group placed Vanessa in the centre of the panel or photo so it couldn't happen again. It was an illustration of sneaky racism that permeated our world.

In my next newsletter to clients, I opened with how important it was to stand up and speak out against racism. About ten people unsubscribed including a large and regular direct client. They were clients we didn't want, and my words seeded action.

My sister, Foo, read a chapter from one of her books, about her experience with racism in apartheid South Africa. Decades before, a colleague of hers, a black empowerment star, smiled through years of racist jokes by the conservative white men in the company. She and he were close friends, and they stood up for each other, both different from the rest, she a woman in a male-dominated environment, and he, black in a white-dominated environment. Foo lost touch with him when she moved to Canada. She later learned he had drunk himself to death. Racism had crushed him. They didn't put their knee on his neck, but decades of undermining jokes had had the same effect.

Foo told me that my newsletter was the encouragement she needed to speak out and read the chapter to her followers. She had a much larger mailing list than we did. It reinforced how important it was to act, your actions lead to actions beyond you. The more of us that speak out, the faster the change.

Each day I lay down on the spring grass like I had when the corks disappeared in the Corona vortex. It brought bliss. I was one with the earth. Pure heaven was right here and right now, ions bouncing through my body.

Lockdown gave us time to think and plan. Like my personal audit, we did a farm audit to assess where we could reduce our carbon dioxide emissions and improve our environmental actions. Glass bottles are often the biggest generator of carbon dioxide for a winery business. We calculated that one hectare of our forest area consumed the equivalent of carbon dioxide generated by the glass bottles used at Château Feely. We already used 'ecova' bottles that were engineered for energy efficiency in production, and weighed less, on most of the range. After doing the audit we changed all of the range to ecova. We kept pushing our suppliers to provide better solutions. After years of asking for it, our label supplier offered us 100 percent recycled paper. We realised that we could encourage change through our business relationships too.

We looked for ways to close the circle in every part of our business. When we delivered a new order to local restaurant clients, we collected the used cardboard boxes. When the boxes were too worn to reuse, we used them in the vegetable garden. We composted all our living waste. We said no to unnecessary packaging. We assessed our energy use and our water use. Doing the audit felt good, it helped to see where we could pat ourselves on the back but also where there was room for improvement.

Lockdown ended and summer guests poured in. Visits and tastings had to be outdoors to avoid spreading the virus. Elodie and I christened the gravel between the tasting room and the grassy area under the trees 'the burning desert.' We trekked back and forth from fridge to clients feeling slightly surreal. The heat created a sense of floating, shimmering like a mirage, our masks wet with sweat.

An email from Magendie announced that Ellie had been accepted. From September both our daughters would be at high school in Bordeaux. Sean and I would be phase one empty nesters.

CHAPTER 18

Wasps in the empty nest

The sunrise met me as I descended the hill, foraging for blackberries. Aromas of cut grass and forest floor mingled in the cool air. At the bottom of the vineyard, I followed the forest line, lifting my hand to run along the trees and brush, sensing the textures of the different leaves, their calmness, and their vibrant greens. I felt like they were talking to me, like the oaks on Sherborne Hill, when I visited the UK and encountered Stonehenge for the first time.

Sunrise was the best time to be out before the heat layered up waves of shimmer across the vineyard and everything went silent. The forecast was for 39 degrees Celsius. When it was that hot, I understood what David Wallace-Wells, an American journalist known for his writings on climate change, meant in his book 'Uninhabitable Earth'. We were making it uninhabitable for ourselves, but also for all the other creatures. I picked blackberries, one for me, succulent and juicy, one for the bowl, sometimes two for me, one for the bowl. It filled slowly with my belly.

Later, the coolth of my morning berry hunt was long gone. It was hot. Extremely hot. Relieved to be back from the walking part of the tour, I indicated to our clients to settle at the table under the trees and crossed the burning desert to the tasting room door.

Sophia flung it open.

'You've got to go and see Papa now. He's on my bed. He's been stung. He thinks he's going into anaphylactic shock.'

I sprinted across the courtyard to our house. Sean was lying on the bed reading on the iPad. His face was red and puffy, and his arms were in constant movement, reaching up to scratch, and being forced back down to his sides, as he tried to control himself.

'I'm fine,' he said. 'But I think I'm going into anaphylactic shock. It says we should do something.'

I had never seen anaphylactic shock. Sean had been stung many times with little repercussion. My brain tried to process what was going on and what to do. I recollected reading that pharmacies could supply a solution to allergic reactions.

'We need to get you to the pharmacy,' I said. 'Let's go.'

I grabbed the car keys and ran to the tasting room for my bag and phone. I found my bag but not my phone.

'Where's my phone?' I demanded.

'Sophia had it,' said Ellie.

We all looked around.

'Quickly, I need to go. This is a crisis.'

'It's in your hand,' said Sophia, pointing.

Panic had a strange effect on the brain.

It was Elodie's day off. I asked Sophia to take glasses, water, wine, and an opener, to the clients. At under 18, she couldn't serve them, but they could serve themselves.

Sean staggered across the courtyard to the car. I pointed to him as I shouted, 'I'm so sorry, I have to go. My husband's going into anaphylactic shock.'

'Go, go!' they yelled.

On the descent from Saussignac, the limit was 80 kilometres per hour. It should have been 40. I was going 60, fast for the road, but not for the circumstances.

'You're going too fast. Slow down Caro,' said Sean.

I slowed a little. After crossing the D14, on the stretch to Gardonne, a flat wide country road, I picked up the pace.

'You're going too fast. Way too fast. I'm okay. Slow down,' said Sean.

I could see him swelling. He was not okay.

We skidded into the pharmacy car park. I grabbed the two masks in my bag, thank God they were there, as I hadn't thought of packing them, and passed Sean the flowery one that I hadn't used for a couple of days. We ran inside, Sean a pace behind me.

I jumped the queue, shouting '*Mes excuses, c'est une urgence.*' My apologies, it's an emergency.

The owner and chief pharmacist, one of many behind the counters, came forward.

'Please help us,' I said. 'My husband is going into anaphylactic shock from a wasp sting.'

'How long ago?' he asked.

'About an hour and a half,' replied Sean.

I saw a look pass his face that said, 'Idiots, why did they take so long to react?'

I felt like saying the same thing.

'*C'est trop tard*,' he said. 'It's too late.'

My heart raced off the charts.

'You have to go to emergency in Ste Foy. What I have won't work at this late stage.'

We leapt back into the car. Outside the town limit, I accelerated.

'You're going too fast,' said Sean. 'I'm okay.'

I looked across at him. His face was blotchy, his lips puffed out like they had had a Botox overdose, and his hands rubbed his arms like someone possessed. I ignored his demand and kept the pace, as I interrogated him about the long time between the sting and informing me.

'I thought it was a touch of heat stroke, so I took a cold shower and lay down. When it got worse, I started reading up and realised it was a reaction to the sting and could be serious. That's when I asked Sophia to go and get you. Slow down.'

Sean's anxiety about my speed, which was never more than ten kilometres above the limit, grew as he expanded. We read later that anxiety is part of the reaction.

He started to wheeze.

'Seriously. I'm' -wheeze- 'fine.' Wheeze. 'You can slow down.' Worse wheeze.

I skated around the traffic circle at the entrance to Ste Foy feeling a sense of déjà vu. The guy on the access boom at the hospital took one look and opened the gate. There was no need for an explanation.

We grabbed our masks. At the emergency entrance, I pressed the buzzer and explained. A wiry, young man came out within seconds.

'Come with me Monsieur,' he said, then threw back over his shoulder to me, 'I'll be back for his details.'

The door closed behind them. I stepped out of the entrance alcove and onto the concrete ramp that descended to the car park. There was no waiting room access because of the pandemic. Four other people were interspersed along the ramp waiting for news. I descended to the tarmac and called Sophia to give her an update and check what had happened with the abandoned clients. She had gone inside to change, and they had left by the time she came back out. Why she needed to change was something only a teenager could answer. Fortunately they hadn't prepaid.

On the hospital tarmac, the midday sun did its best to give us heat stroke. The receptionist nurse called me. He took me into the cool interior to gather Sean's details. It didn't take long since all the Emergency Rooms in the vicinity of Château Feely had Sean's information from previous visits.

'You can wait outside. It'll be at least two hours,' he said.

'If it's going to be that long I'll go home and wait to hear from you,' I said. Staying on the hot tarmac for two hours in the heat of the day was a recipe for another emergency patient.

'Okay,' he replied and took my mobile number.

Driving away, I felt calm. It felt safe leaving Sean there, a testament to our experience with French hospitals when the chips were down. I trusted them. But as the afternoon stretched beyond the promised two hours my anxiety rose. I felt like I was in a twilight zone. Eventually, I phoned them. Sean was progressing well, but it would be another couple of hours. Relief. Even more so when Sean called us in the late afternoon and we could hear his voice. They needed to observe him a bit longer. As evening settled in Sean called again to say that he might need to stay overnight. They wanted to observe his heart.

It felt odd eating dinner without Sean. As we finished cleaning up, he called to say I could fetch him. Back in Saussignac, we gathered around the kitchen table to hear the full story.

'The second I arrived the doctor and five interns and nurses raced around injecting me with stuff, getting IVs into me. They moved so fast.'

'Wow. Like in the movies!' said Ellie.

'Yep,' replied Sean. 'They were worried. When the doctor debriefed me, he said my throat was closing. We were lucky we got there when we did.'

I felt a shot of adrenaline. It had been that close. Thank God I ignored his demands to slow down.

❧❧❧ ❧❧❧

Harvest started a month earlier than our first harvest. The climate crisis was showing itself full frontal. In Burgundy, where written records of harvest go back to 1354, they declared the earliest harvest since 1500.

We didn't have a chance to rest and breathe before we had to dive in. We hadn't taken a holiday. We were still in the thick of the tourist season. It felt too fast, too brutal, like the summer we experienced. I wasn't ready. But once I contacted people that had harvested before, or might know someone who wanted to harvest, the team fell into place.

Vincent had been picking with us for about six years. He was like a peaceful breath, gentle but strong, a fast picker with a quiet sense of humour. Inès, niece of our neighbours at La Queyssie, on work experience from her hospitality degree in Paris, was serious and reliable. Barak was a mixed martial arts specialist and tattoo artist married to a Saussignac local. He had lived in Saussignac for two years, but we hadn't met before. Handpicking brought people together. He was fast, fit, and fun. Working with him I knew that any grapes I missed would be got. Elodie completed the team. She had passed her diploma and was due to leave us a month after harvest. We'd miss her. She was funny, great with people and gifted in many ways including video editing, something that had helped save the year for us. Given the pandemic, we were in less of a position to hire a permanent employee than before.

It was our smallest team ever, partly for budget reasons and partly to decrease the Covid risk. Usually, we worked in teams of two, spaced a metre apart, now we spaced wider, and instead of switching duos daily, we kept our pairs for the whole harvest. Like the previous year, the picking was so early that Sophia and Ellie were still on holiday. We started on a flower day on a descending moon on the biodynamic calendar, with Merlot grapes for a dry rosé. The conditions were perfect, and the delicate pink juice was fragrant with strawberry, citrus, and quince. From there, our *'petite équipe de choc'* – small, powerful team – took on the Sauvignon Blanc. The juice was rich, hints of gooseberry, mango and passion fruit, a different aroma range to our usual Sauvignon Blanc. The heat of the year brought a more tropical aroma palette.

Between the Sauvignon and the Semillon, I took Sophia and Ellie to school in Bordeaux. The first Covid lockdown had run from March to the start of summer holidays giving our daughters nearly five months at home. This new school year we faced a significant change in our lives, both daughters in boarding school, and Sean and I in an empty nest.

Bordeaux's traffic felt harassed, crowded, rowdy. I had lost the knack of driving in the city. After dropping Ellie for her initiation, Sophia and I walked the streets. It felt weird to hang loose with no grape stress.

The hostel had a strong institutional whiff of chlorine and discipline. The speech for new arrivals reinforced how peculiar this start to the school year was. Masks were mandatory everywhere in the hostel. They could be removed at your desk, in bed in the dormitory, or when seated to eat in the canteen. Meals were limited to short windows for restricted groups, making mealtimes even more pressured than they had been. Students had to bring their own water bottles as water jugs would not be on the tables anymore. Ellie said when she and her new classmates took their masks off at lunch it was a 'face reveal', seeing each other for the first time after spending the morning together. It was a bizarre way to start her new life.

I expected to be depressed about leaving my daughters, but instead, I felt excited for their future in this excellent school in beautiful Bordeaux. Ellie loved the first day. She dove into meeting and exchanging with her new tribe, leaving her shyness behind. Sophia was excited to be back with her friends. It was a positive start. They didn't enjoy the rules of the hostel, but it provided a safe place to sleep and eat, and to be with other people on the international programme.

I had harvest on my mind and Ellie had dorm mates to get to know. Sophia's dorm was one door down from hers. Leaving didn't feel difficult at all. This empty nest thing was a piece of cake. As I took off, I felt nostalgic as memories of their primary years flashed in my brain. My GPS saved me from melancholy by telling me to turn into an oncoming tram. Traffic packed in behind me. I waved frantically and backed out of the tram's way hoping the glowering drivers would see the Dordogne registration and forgive a country bumpkin. My heart racing, I edged out of the city.

The next morning the grapes were luscious and gold in the morning light. On my first row, I came upon an old vine with long-stemmed, almond-shaped grapes, obviously not Semillon. I crunched into one and the flavour of lychee exploded in my mouth, leaving a long tropical finish. Something about the year, the vine, and where it was, had brought it out. There was backache, but also joy in this act of handpicking. By the end of the next day, the grapes for the future Feely *Générosité* barrel-aged dry white, a long aging wine, perfect for autumn and winter dining, were in the winery.

Jamie, a friend of friends, joined us for a couple of weeks to learn the winery ropes from Sean. Nightly discussions around the kitchen table were long and deep, ranging from the intricacies of natural fermentations to world peace. We finished harvest earlier than ever before. Global warming was advancing fast. Despite the speed and intensity that the heat created, we loved that intimate harvest.

Mid-week, an email from the school said the hostel would be closed for a week as one of the boarding house staff was suspected to have Covid. The hostel students were kicked out onto the street with no notice. The world had gone insane. Sophia and Ellie took the next train home. Luckily there was one since the pandemic meant fewer trains were running.

We felt blessed by that unexpected Friday to share with our daughters. I didn't know what we would do for their accommodation the following week. Some classmates offered to host them, but expecting people we didn't know, to host our kids for a week, during a pandemic, felt like a big ask.

That gifted day, Sophia pushed the boundaries. She asked to go and stay with a friend that night. It was a boyfriend she had assured us was 'just a friend', but now it was more than that. We discussed all the aspects of staying over, and what it meant. I felt like it wasn't a good idea. On the other hand, she was only a few months from turning eighteen, and having full adult rights.

As we debated the question, I felt like I was under a steamroller, like I was trying to put my ideas above hers, and failing. I stayed over at my boyfriend's house when I was her age, but they had a spare room for me. I felt like I was caught between two generations, like I could feel the influence of my parents and my Protestant upbringing on my feelings about it, and at the same time the powerful influence of Sophia and her decision, her strong will to push it through.

We hadn't met the young man. The summer and harvest frenzy had blinded me to the progression of her relationship. They had known each other vaguely for two years but had connected at a party in the summer. In the weeks that followed, Sophia went into Bordeaux for the day, to 'see friends' a few times. It was to see him, but she wasn't ready to tell us.

Now she was ready, but we weren't ready. Saying no was saying we didn't trust her judgment. I was concerned about potential risks. We interrogated ourselves, talked about it with Sophia, then went our separate ways to work for the afternoon with no final decision made.

We reconvened in the kitchen late afternoon. It felt so hard like my head was splitting in two. I had my parents' mores on one side and my daughter on the other. Sophia reminded us that she was a mature and responsible person. This was her life, her decision. Us saying no would offer reason to go behind our backs, it would say we didn't trust her judgment.

That night, when I ran her down to the station to catch the train, it felt like a real parting. As I drove back up to Saussignac, I had heaviness in my chest. This was what an empty nest felt like. It was not a piece of cake. I woke at 6 a.m. feeling hungover despite having drunk nothing the night before. My shoulders were tense. There had been no call in the night. Sophia had passed through it safely. I came down to find Sean already in the kitchen.

I put the kettle on. Sean turned the speaker onto a random Spotify list. A song called '*Je Vole*', sung by French singer and actress, Louane, about taking flight as a young adult filled the room. Our eyes met across the space, and we burst out laughing. This was real parenting. We were learning to let go.

⤜⤜⤜ ⤛⤛⤛

The last of the cranes flew south with the first frosts. Sometimes the group would hit a thermal and float in spirals for a while before continuing. Even on their great journey, they took moments to rest. It was a message for me. The vine leaves fell. In the clear crisp night sky, Venus and Jupiter were as close to each other as they had been when Galileo observed them cosying up in the night sky in 1620.

The following summer the pandemic eased but the climate crisis cranked up the ante. Europe experienced record wildfires, floods, and temperatures. Our planet was hitting the emergency button and yet, at COP26, the proposals laid out by participants, and in the overall agreement, put us on a path to well over 2 degrees Celsius by 2050. In addition, as with all preceding COPs, nothing was binding.

A study shortly before showed that people were convinced that we humans were causing climate change and that it was real, but less than half of those people felt it would affect them directly. There is denial about how connected we are, and how critical the life support systems of earth are to us. We have become disconnected from them in closed houses and offices, believing ourselves more powerful than nature.

Directly experiencing a hurricane, a flood, a wildfire, an extreme heatwave, or any other catastrophic results of climate crisis, quickly dispels that idea. The truth is, before any fancy comforts, we need air, water, and food. These life-critical elements are being damaged by burning fossil fuels and destroying biodiversity.

This universal connection to everything and every place is true for the climate crisis. My carbon dioxide emissions contribute to climate change across the world, for example, floods in Mozambique, fires in Australia, heatwaves in Dubai, and hurricanes in the USA.

The radio announced that carbon dioxide emissions were set to hit a record high. The next story covered a 30 percent rise in prices for *matières premières agricoles,* base agricultural products, like wheat and milk. The interviewee said it was due to crop failure, resulting from climate change. The interviewer did not make the connection to the previous story about carbon dioxide and how this was directly linked and that each of us was responsible and could act. The crop failure was not a one-off bad year. It was the climate crisis created by greenhouse gases. We need to make that connection constantly, so reducing carbon dioxide emissions remains top of mind for all of us, in all our decisions both long and short-term. The media was not doing that.

Despite the dramatic reduction of air and road traffic during the pandemic, we were still accelerating in the wrong direction. Carbon dioxide recorded at Mauna Loa Observatory in Hawaii increased from a monthly average of 410.48 ppm in November 2019, around the time of COP25, to a monthly average of 415.01 ppm for November 2021, around the time of COP26. On 30th December, mid-winter, I pulled wood off the vineyard trellises dressed in a bikini top in 20 degrees Celsius, another temperature record.

In May 2022, the carbon dioxide level in the atmosphere was the highest ever recorded, at 421 ppm. It was as high as it was more than 4 million years before when seas were up to 25 metres higher, a level that would put many of the world's largest cities underwater. For every one of us that has the necessities of food, water, and shelter met, stopping burning fossil fuels has to be our most important objective, followed by saving biodiversity. In our modern digitalised world, we feel less connected to nature, but the effects will trickle down. We are dependent on the gifts of nature, and the earth, for our air, water, food, and materials. Everything, every action, every decision, and every purchase, must consider the carbon dioxide and biodiversity impact. It's hard. But our lives depend on it.

In June 2022, we were hit by a devastating hailstorm that wiped out 70 percent of our harvest. The hailstones were like golf balls. They left the vines and the potager vegetable garden in tatters, a windowpane broken, and our car, like it had been hammered by a demented giant. The hail scythed canes, flowers, and tiny grapes off the vines. The thundering of the stones on the roof was like a battle overhead. We watched, powerless. As the sun rose, we walked the vineyard to assess the damage. The ground was white with hail. The vines were badly hurt. I could hear them crying. I felt traumatised like they were.

We were in a fog that day, like a day of mourning. The vines swift regeneration told us to pick ourselves up. Within a few days their wounds had closed, and healing was well underway. Ten days later, new stems and leaves were forming. They wouldn't bring back the lost harvest, but they would help to ripen what was left, and repair and form potential canes for next year.

With regenerative farming, our vines are more resistant to climate extremes[17]. Not only that, through our farm we share the story of regenerative agriculture, of organic farming, and how they can help mitigate and adapt to the climate change we see first-hand. Farms can offer a place to discover nature, they can connect people to the earth via their approach and outreach. For example, the team at Domaine Montirius in the Southern Rhone, whom I met at the Biodynamic conference in Alsace, take a moment every morning to check in and be thankful for the day, a simple gesture that offers respect to themselves and a moment of gratitude to the earth. Taking their lead, as part of my meditation after yoga every morning, I include a moment of gratitude, for humanity, plants, trees, and creatures great and small, for this incredible globe riding through space, our home.

We are living climate crisis and unless we all change our habits it is going to get much worse. Outsize cars and private jets are still seen as 'cool' despite not being okay for our collective future. For some, vine rows with nothing growing underneath them, look neat. I see a desert, and cancer if herbicide was used. Our perceptions are still being formed by images that seemed okay in the 20th century but are not okay in the 21st, given what we know about the climate crisis and biodiversity loss. The tide is turning. We contribute to this change by our actions, by what we support, and by what we share and like on social media.

The weight of climate change is heavy. I lie down on the grass and breathe. Almond leaves flutter against a blue sky. I remind myself that the sun is shining and beyond it, the sky is full of stars. We are voyaging on a small blue dot in an infinite universe. I feel deep joy and care for the earth beneath me. Gratitude, and peace, envelope me.

The threat of devastation will not solve the climate crisis or loss of biodiversity. Feeling estranged and threatened will make us less able to transform. After the hail, the vines' regeneration was so swift and miraculous, it gave me goosebumps. They offered hope and awe at their capacity to regenerate.

Living light on the planet connected to a thriving local community with access to nature offers more happiness and is more likely to succeed in addressing these challenges than changing SUVs from fossil fuel to electric, or blue-sky carbon capture solutions. The SUV on electric has swapped drilling oil for mining precious minerals, and the environmental devastation that that brings. Once we have burnt the carbon dioxide into the atmosphere we can't get it back, and there is no certainty that we ever will, so we are better off not burning it in the first place.

As the happiness project at Harvard[18] proves, happiness is not increased by more money and more stuff. For those of us with our basic needs met, we have to choose an ecological lifestyle or risk losing the possibility of life at all. We are in this together.

This 'we' goes beyond us, to include the Earth and everything that is part of it. Robin Wall Kimmerer's beautiful book, 'Braiding Sweetgrass', tells us that when we see other creatures, plants, and even rocks, rivers, and mountains, as beings, we appreciate and respect them, and then we treat them with more care. In the process of learning this way of looking wider and deeper, we transform ourselves and have the energy and insight to transform our way of living. I find this message, and great solace, in Thich Nhat Hahn's 'Zen and the Art of Saving the Planet'.

Humans have the capacity to pull together in a crisis as long as we feel common cause. I hold hope for a more spiritual, holistic approach to living our individual and collective lives as we move into a post-industrial age. I see power in meditation, yoga, skill circles, gardening, activist movements, regenerative agriculture, kindness, and celebrating what brings us together and cultivates change. We are in a 'climate crisis'; but crisis also means opportunity for transformation. We can cultivate change that leaves us, and our world, in a better place than it is now.

Yoga has helped me to include more feeling and thought in my life, as flagged by the biodynamics conference in Alsace, and brought me a sense of contentment I never had before. Sean's permaculture garden and agroforestry in the vineyard are helping our farm weather the climate crisis and creating more biodiversity. Farming in climate change is tough, but working close to nature is a gift, one we are deeply grateful for every day.

Everywhere, people are starting to change and cultivate change. My friend Lee, in Australia, found a new job with a company that is not destroying biodiversity or creating climate collapse. I am even more determined to make positive environmental change through action, workshops, and books. It is not easy to face the facts of climate change and biodiversity loss or to change our lifestyles, but we can do it. We can be brave.

With our daughters in Bordeaux, Sean and I have time together that we haven't had since they were born. We are rediscovering each other. Raising a family is worth every challenge. I rise from the grass. It has made a beautiful pattern on my inside wrist. Maybe it's time to get a tattoo.

1. https://ijsrset.com/paper/2752.pdf

2. https://news.harvard.edu/gazette/story/2017/04/over-nearly-80-years-harvard-study-has-been-showing-how-to-live-a-healthy-and-happy-life/

3. https://www.acbio.org.za/wp-content/uploads/2015/02/ACB_Open-letter_Monsanto_crop_failure_November-2009.pdf

4. https://www.reuters.com/article/rbssIndustryMaterialsUtilitiesNews/idUSN0220100420090402

5. https://ehjournal.biomedcentral.com/articles/10.1186/s12940-016-0117-0

6. https://www.iarc.who.int/wp-content/uploads/2018/07/MonographVolume112-1.pdf

7. https://www.researchgate.net/publication/24095661_Inequality_And_Crime

8. https://data.oecd.org/inequality/income-inequality.htm

9. Data from the Global Monitoring Laboratory by National Oceanic and Atmospheric Administration (NOAA) of the United States of America

10. https://www.greenpeace.org/international/story/58753/10-things-know-about-ipcc-climate-science-report/

11. http://www.cesnet.co.za/assets/01Wild%20Coast%20Abalone%20-%20Draft%20Scoping%20Report%20-%20December2020(1).pdf

12. IPBES (2019) most recent version at the time of writing: Global assessment report on biodiversity and ecosystem services of the Intergovernmental Science-Policy Platform on Biodiversity and Ecosystem Services. E. S. Brondizio, J. Settele, S. Díaz, and H. T. Ngo (editors). IPBES secretariat, Bonn, Germany. 1148 pages. https://doi.org/10.5281/zenodo.3831673 OR for a shorter read IPBES (2019): Summary for policymakers of the global assessment report on biodiversity and ecosystem services of the Intergovernmental Science-Policy Platform on Biodiversity and Ecosystem Services 56 pages. https://doi.org/10.5281/zenodo.3553579

13. https://www.ncbi.nlm.nih.gov/pmc/articles/PMC8622992/

14. https://www.theguardian.com/business/2018/sep/25/monsanto-dewayne-johnson-cancer-verdict

15. https://youtu.be/tN-LJ7w5pwQ

16. https://youtu.be/7BxQLITdOOc

17. For example, keeping the ground covered in plants helps against flooding and erosion, agroforestry helps cope with late frost

18. https://news.harvard.edu/gazette/story/2017/04/ov
    er-nearly-80-years-harvard-study-has-been-showing-
    how-to-live-a-healthy-and-happy-life/

What can we do? Thrive and Cultivate Change

If this book has opened your eyes to the challenges of climate change and biodiversity loss, but you feel like don't understand them enough, or know what to do about them, join my book club 'Cultivating Change' where we'll explore books that help understand and take action. Crisis creates opportunities for change that can create a better world, one that is kinder, less stressed, and healthier.

Seeking to understand climate change and biodiversity loss and how serious they are, helped me see what I can do and why. I read books about biodiversity and about the climate crisis. Here is a selection of my favourites (find an up-to-date list at https://chateaufeely.com/climate-crisis-best-books/).

Books to learn about climate change and what we can do:

• Being the Change: Live Well and Spark a Climate Revolution by Peter Kalmus
• This Changes Everything: Capitalism vs. The Climate by Noami Klein
• Zen and the Art of Saving the Planet by Thich Nhat Hanh

Books on biodiversity loss (and a little about climate since they are linked)
• Braiding Sweetgrass by Robin Wall Kimmerer
• The Sixth Extinction: An Unnatural History by Elizabeth Kolbert

Examples of movements to participate in – to get your momentum up:
Greenpeace, 350.org, Extinction Rebellion
Take the Jump: https://takethejump.org/
Rewilding your garden (or your window box) to help recover biodiversity: https://wearetheark.org/

Join my Cultivating Change book club to read books like these with other people and to explore the themes in more depth https://carofeely.com/sign-up .

A letter to you dear reader

Thank you for reading this book. The process of writing helped me understand more about relationships, climate change, and what it means to live a good life. If you enjoyed reading it please consider posting a review on Amazon or Goodreads. Reviews make a significant difference to the visibility of a book. Thank you!

I'd love you to share a photo of you and my book (you can do it with the front cover showing on an e-reader not only with a physical copy) with me on social media: https://www.instagram.com/carofeely or https://www.facebook.com/caro.feely.wines

I invite you to join my mailing list https://carofeely.com/sign-up to stay up to date with the latest news about my books. Consider joining my book club Cultivating Change where we delve into books that inform and enchant.

If you are part of a book club already, page over for a list of book club discussion questions. I would be delighted to participate via Zoom or WhatsApp. We can ship Feely organic wine to many countries so you could even have a wine tasting as part of your book club meeting.

You can find out more about Feely organic farm, an award-winning estate with walking tours, yoga (https://chateaufeely.com/) and French Wine Adventures wine school (https://frenchwineadventures.com).

We would love to see you here or online.

With best wishes and gratitude,

Caro

Book club questions

1. Did any part of this book strike a strong emotion in you? Which emotion and which part?

2. Are there any quotes, passages, or scenes you found particularly thought-provoking?

3. What direct impacts of climate change does Caro describe in the book? Have you experienced these? Can you think of any impacts of climate change in your life? Are there any that you can see affecting you directly in the future?

4. How do you think we can solve the climate crisis? What systems need to change?

5. Do you feel like you understand what biodiversity means? How do you think loss of biodiversity affects your life? Or could affect your life in the future?

6. Is this book different from the books you usually read? How does this book compare to other books you've read in your book club?

7. Did you learn something you didn't know before? Has your behaviour changed?

Acknowledgments

A memoir is a personal journey, a way of making sense of the past, so my biggest thank you goes to my family, Sean, Sophia and Ellie, for their participation in this adventure and for agreeing to me sharing this intimate portrait of a moment in our lives. Sean jokingly says I write 'great fiction', which I take as high praise from an ex-journalist who reads literature for breakfast. A special thank you to my beta readers Dave Smith, Jacqui Brown, Deborah Stevenson, Tora Shand, Melissa Pleighty, Garry Mason and Derek and Helen Melser.

Heartfelt thanks to friends and family that are part of this book Ad and Lijda van Sorgen, Sébastien and Véronique Bouché, Steph and Dave Pons, my dear friend Russ, my parents Lyn and Cliff Wardle, Sean's Dad, John Feely, my sister Foo and brother Garth, Bruce, Glynis, and Duncan Bristow, Aideen Dunne, Lee Albert, Thierry and Isabelle Daulhiac, my yoga teacher Sunshine Ross. Thank you to the wonderful people that have contributed to Feely farm, Rosalyn Bowles, Elodie Passera, Duncan,

Jamie, Roddy and Cathy Pugh, Martin Walker and many more.

Thank you to my book cover designer Julie Adams and to my ARC readers that provided input. To you and all of the people that have helped get the word out about this book with social media, blog posts and reviews, I send infinite gratitude.

Grape Expectations

Book 1 of The Vineyard Series.

What does it take to follow your dreams?

The wine filled my mouth with plum and blackberry sensations. A picture of a vineyard drenched in sunlight formed in my mind. Sean drew me rudely back to our small suburban home.

'How can they be in liquidation if they make wine this good?'

When Caro and Sean find the perfect vineyard near Bordeaux their dreams of a new life in France are about to come true. They arrive, with a toddler and a newborn, to face a dilapidated farmhouse, and challenges including accidents with agricultural equipment, cultural misunderstandings, and money worries. Undeterred, they embark on the biggest adventure of their lives – learning to make wine from the roots up.

Reviews

'A beautifully written tale of passion and guts.' Alice Feiring, Author

'Captivating reading' Destination France

'An inspiring story of how one couple changed their lives.' Jamie Ivey, Author

'I was delighted by this book, by what it says about the passion of winemaking, France, family life, and the challenges that build a marriage.' Martin Walker, bestselling author of the Bruno, Chief of Police series

E-book ISBN: 978-2-9586304-1-6
Print Book ISBN: 978-2-9586304-0-9
By Caro Feely

Saving Our Skins

Sequel to Grape Expectations. Book 2 of The Vineyard Series.

'Earnest and winning... sincere and passionate'
The New York Times

'We have to get to the next level, or we have to get out,' I said.

'We have to have more vines and more accommodation,' replied Sean.

Both would take investment we didn't have.

For Caro and Sean, building their vineyard dream and overcoming challenges that include a devastating frost, bureaucracy, and renovation setbacks, will take courage, ingenuity, and luck.

This book is about love and taking risks while transforming a piece of land into a flourishing organic vineyard and making a new life in France. It explores the reality of following your dream, challenges of building a new business, renovating in France and raising a family as a working mum; and includes nature, delicious food and wine, and voyages to Napa and Sonoma wine regions in the USA, and to Alsace, Bergerac, Bordeaux, and Burgundy wine regions in France.

Reviews

'So impassioned that it could inspire you to drop all security, move to the backwaters of France, and bet your life, all for the love of making wine.' Alice Feiring, author and wine writer

E-book ISBN: 978-2-9586304-3-0
Print book ISBN: 978-2-9586304-2-3
By Caro Feely
First published in 2014.

Vineyard Confessions

Book 3 of The Vineyard Series. Initially published under the title 'Glass Half Full' in 2017.

How do you balance a growing business and family life? Is it possible to have it all?

'Hand harvesting was different to machine harvesting. It was convivial and slow. We started at dawn and proceeded across the vineyards. It was better for us and for the grapes, the human scale and pace of it was peaceful and joyful. It gave us time to share confidences and confessions.'

But this rose-tinted glimpse of Sean and Caro's French vineyard life is only part of the story – with it come long hours and uncertainty. The rollercoaster ride of managing a growing business and navigating menopause is as challenging as making natural wine in harmony with the environment.

In this book you will discover the joys and challenges of living your dreams; navigating life changes, why organic matters and what organic wine, biodynamic wine and natural wine are.

Join Caro on her search for balance in love and wine.

Reviews

'A love story poured beautifully.' Robyn O'Brien, bestselling author

'Caro Feely is a force of nature! Caro draws the reader into her world with its all of its challenges, triumphs and heartaches. Required reading for winelovers everywhere.' Mike Veseth, author of Wine Wars and The Wine Economist blog

'Honest and touching. Caro Feely gives us the real thing including why we need to heal our soil and change the way we farm.' Martin Walker, bestselling author

'A brave and compelling tale' Alice Feiring, author and journalist

E-book ISBN: 978-2-9586304-5-4
Print book ISBN: 978-2-9586304-4-7
By Caro Feely
Retitled, edited and republished in 2023.

Saving Sophia

"The paediatrician came back without our baby.
'I'm afraid we have a problem.'
Stars bounced around my vision of the doctor's talking head, his voice far away like a dream. I felt like I was sinking into a void and leant back to steady myself on the bed. We had experienced the deepest, most life affirming event the night before, the birth of our first child. Now she was being taken away. It was a moment that would take us on an intense journey and change the course of our lives."

'Saving Sophia' is a page-turning true-life story of pregnancy, birth, ICU, breastfeeding, first time parenting, postnatal depression, and raising a family. A gripping and humorous account of a nail-biting entry into parenting that offers a glimpse into, and a reminder of, how miraculous raising a family is.

This is a memoir with love at its heart, a story of first-time mothering, family, and friendship.

Key themes

Primary themes: motherhood, ICU baby, breastfeeding, first time parenting.

Secondary themes: healthy living, yoga, belonging, immigration, stop pesticides, stop glyphosate, postnatal depression/ postpartum depression, follow your dream.

Reviews

'I loved this wonderful book.' Julie Haigh, We Love Memoirs

'A heart-warming and deeply moving memoir of love and family.' Martin Walker, bestselling author

'A beautifully written memoir... I couldn't put it down.' Jacqui Brown, Book blogger

'What a read! Unputdownable,' Jeanne Wissing

E-book ISBN: 978-2-9586304-9-2
Print book ISBN: 978-2-9586304-8-5
By Caro Feely
Publication 5 November 2023.
A prequel to the vineyard series.

Printed in Great Britain
by Amazon